The Wonders of Diptera - Characteristics, Diversity, and Significance for the World's Ecosystems

Edited by Farzana Khan Perveen

Published in London, United Kingdom

IntechOpen

Supporting open minds since 2005

The Wonders of Diptera - Characteristics, Diversity, and Significance for the World's Ecosystems
http://dx.doi.org/10.5772/intechopen.91609
Edited by Farzana Khan Perveen

Contributors
Darine Slama, Emna Chaker, Hamouda Babba, Manoel Uchoa, Nádia Roque, Morgana F. Wachter-Serapião, Murat Helvaci, António Souza, Eduardo Jose De Arruda, Alex Martins Machado, Taiana Gabriela Barbosa De Souza, Raphael Antônio Borges Gomes, Sevidzem Lendzele, Ovono Mélodie Audrey Prisca, Mounioko Franck, Zinga Koumba Christophe Roland, Maroundou Audrey Pamela, Acapovi-Yao Géneviève Lydie, Tamesse Joseph Lebel, Simo Gustave, M'batchi Bertrand, Mavoungou Jacques François, Karima Zerguine, Azubuike Christian Ukubuiwe, Israel Kayode Olayemi, Chinenye Catherine Ukubuiwe, Bright Ugbede Sule, Rajendra S. Fartyal, Pragya Topal, Divita Garg, Jimi N. Nakajima, Anderson S. Fernandes, Farzana Khan Perveen, Anzela Khan

First published in London, United Kingdom, 2021 by IntechOpen
IntechOpen is the global imprint of INTECHOPEN LIMITED, registered in England and Wales, registration number: 11086078, 5 Princes Gate Court, London, SW7 2QJ, United Kingdom
Printed in Croatia

British Library Cataloguing-in-Publication Data
A catalogue record for this book is available from the British Library

Additional hard and PDF copies can be obtained from orders@intechopen.com

The Wonders of Diptera - Characteristics, Diversity, and Significance for the World's Ecosystems
Edited by Farzana Khan Perveen
p. cm.
Print ISBN 978-1-83968-882-9
Online ISBN 978-1-83968-883-6
eBook (PDF) ISBN 978-1-83968-884-3

For EU product safety concerns:
IN TECH d.o.o., Prolaz Marije Krucifikse Kozulić 3, 51000 Rijeka, Croatia, info@intechopen.com or visit our website at intechopen.com.

Meet the editor

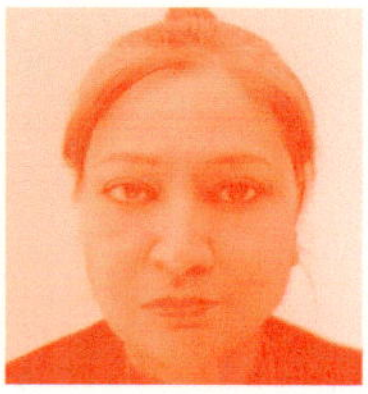

Dr. Farzana Khan Perveen (FLS; Gold Medalist) obtained her BSc (Hons) and MSc in Entomology from the University of Karachi, Pakistan, and MAS (Monbusho Scholarship) in Agronomy from Nagoya University, Japan, and a Ph.D. in Toxicology from the University of Karachi. She is the founder of the Department of Zoology and former controller of examinations at Shaheed Benazir Bhutto University, Hazara University, and Kohat University of Science and Technology. She is the author of 150 high-impact research papers, 135 abstracts, 40 authored books, 9 chapters, and 9 edited books. She is also a student supervisor. Her fields of interest are entomology, toxicology, forensic entomology.

Contents

Preface

This book provides comprehensive and concise knowledge about Diptera, an order of insects that has both useful and harmful aspects for humans, animals, plants, and the environment. Insects of this order act as agricultural pests as well as vectors of diseases and carriers of microorganisms. Control of pests and vectors has become a major issue and crucial factor for future technology that must meet certain requirements to secure the health of humans, animals, and crops. This book presents in-depth knowledge and recent advancement in research.

Section 1: " Introduction to Diptera"

Chapter 1, "Introductory Chapter: Diptera" by Perveen and Khan is an introduction to the order Diptera, an order of single-pair winged insects commonly known as flies or true flies. The chapter also presents the history, classification, external features, economic importance, usefulness, and harmful effects of these insects. It also examines their roles as pests and their medical importance, internal anatomy, and physiology including nervous, digestive, excretory, respiratory, circulatory, endocrine, and female and male reproductive systems.

Chapter 2, "Characteristics of Dipteran Insects" by Helvacı discusses fly metamorphosis. Except for mosquitoes, Dipteran insects have sponging mouthparts. Important examples of Dipteran insects are the olive fruit fly, *Bactrocera oleae* (Rossi) and the Mediterranean fruit fly, *Ceratitis capitata* (Wiedemann), both of which cause damages in agricultural production. *B. oleae* is the major pest in olive production and *C. capitata* cause damages in fruits production.

Chapter 3, "Fruit Flies (*Drosophila spp.*) Collection, Handling, and Maintenance: Field to Laboratory" by Topal et al. discusses Drosophilae, which are versatile, low-maintenance, and non-harming model organisms. They can be easily used in all fields of life sciences like genetics, biotechnology, cancer biology, genomics, reproductive biology, developmental biology, microchemical studies, ecology, and others. This chapter emphasizes techniques of capturing these flies and discusses species-specific baits to catch more yield. The chapter also discusses using culture food media to collect samples and the reasons for using each ingredient in the media. Finally, the chapter highlights basic clues for identifying different species and sexes in the field and lab.

Section 2: "Distinctive Diptera"

Chapter 4, "Diversity of Tephritidae and Agromyzidae (Diptera: Brachycera) in Flower Heads of Asteraceae in the Chaco" by Manoel et al. discusses Asteraceae daisies. There are nearly 24,000 species of Asteraceae herbs and shrubs worldwide that coevolved with several taxa of endophagous insects including Agromyzidae, Ceciidomyidae, Tephritidae, Coleoptera (Apionidae), Hemiptera (Miridae), Lepidoptera (Blastobasidae, Gelechiidae, Pterophoridae, Pyralidae, and Tortricidae), and the parasitoids of these insects. Daisy flower heads provide ideal conditions for food, breeding, and shelter of these insects. The florivorous flies in Brazil belong to the subfamilies Agromyzinae (Agromyzidae) and Tephritinae (Tephritidae). The chapter focuses on these insects, their host plants, and their parasitoids in flower heads of Asteraceae from the Brazilian Chaco.

Chapter 5, "Feeding by Florivorous Flies (Tephritidae and Agromyzidae) in Flower Heads of Neotropical Asteraceae (Asterales) from Central Brazil" by Manoel et al. investigates the occurrence of Tephritinae and Agromyzidae flies associated with flower heads of Asteraceae species in six different phytophisionomies: three in the Dourados region and three in the Brazilian Chaco (at Porto Murtinho), Mato Grosso do Sul. The chapter evaluates 54 species of Asteraceae in these two regions. Larvae of tephritine, depending on the species, consume leaves, stems, flowers, or roots of their hosts. Some tephritines feed on flower heads of the Asteraceae weed and can act in population suppression of invasive species in cultivated areas. Agromyzidae are represented by tiny phytophagous flies.

Chapter 6, "Chironomidae: Biology, Ecology and Systematics" by Zerguine discusses the Chironomidae family, which is a group of Diptera belonging to the suborder of Nematocera, commonly called non-biting midges in the adult stage and bloodworms in the larval stage. Chironomidae are often the most abundant group of macroinvertebrates, in number of species and individuals, encountered in all aquatic environments including freshwater, brackish water, terrestrial water, and even the sea. This family is divided into eleven sub-families that have different ecological status. This chapter examines life cycle, subfamilies, and ecology of Chironomidae.

Chapter 7, "Ecological Aspects of Tabanids (Diptera: Tabanidae) in a Gabonese Cattle Ranch" by Ovono et al. discusses the fight against mechanical vectors of trypanosomosis. It investigates the abundance, species diversity, and daily activity of tabanids in a cattle ranch in Gabon. Nzi and vavoua traps were used to catch 616 tabanids in three divisions of this ranch: 349 (56.66%) in Division 1, 226 (36.69%) in Division 2, and 41 (6.66%) in Division 3. In Division 1, *Tabanus. taeniola* was the most abundant species with an Apparent Density per Trap (ADT) of 2.2, followed by *Haematococcus pluvialis* (ADT = 1.05). In Division 2, *H. pluvialis* was most abundant species with ADT of 1.6, followed by *T. taeniola* (ADT = 0.38). In Division 3, the most abundant species was *H. pluvialis* (ADT = 0.15). There was no statistically significant difference in catches with trapping sites. Division 3 recorded the highest diversity index values. The nzi trap recorded greater tabanid catches than the vavoua trap. The diurnal activity rhythm of the most frequent species encountered differed slightly with prospection sites.

Chapter 8, "Morphological Keys for the Identification of Tunisian *Culicoides* Biting Midges (Diptera:Ceratopogonidae)" by Darine et al. discusses *Culicoides* biting midges, which are tiny blood-feeding insects that carry several diseases with veterinary and public health significance, including bluetongue in ruminants, African horse sickness in equids, and filarial diseases like onchocerciasis and mansonellosis. Their identification depends on the microscopic examination of key morphological characteristics. Consequently, identification keys are important to any non-experiment working with these biting midges. Tunisian fauna of *Culicoides* biting midges consists of thirty-five species, whose morphological delineation may be troublesome for non-taxonomists. In response to this situation, and for the first time, a key to the adult *Culicoides* species in Tunisia was prepared.

Section 3: "Control of Mosquitos"

Chapter 9, "Control Strategy for *Aedes aegypti* (Linnaeus, 1762) Population" by de Souza et al. describes the *Aedes aegypti* (*Culicidae*) mosquito, which has adapted to different environments, mainly urban ones. They have a high degree of vector competence for viral diseases, especially dengue, the arbovirus with the highest number of cases in the world. The adaptive ability of this insect and the abundance of breeding sites have undermined attempts at population control, resulting in a high degree of infestation in many regions of the world and a dengue endemic. It is important to understand the different nuances of the insect to understand the adaptive capacity of this vector in order to propose new strategies of population control, especially in periods of greater proliferation. This chapter discusses population control strategies in different scenarios, mainly in Brazil.

Chapter 10, "Environmental Manipulation: A Potential Tool for Mosquito Vector Control" by Ukubuiwe et al. discusses mosquito-borne diseases. Chemicals are mainly used to control all life stages of mosquitoes. However, an increase in resistance to commonly used insecticides has led to renewed efforts for vector control. Environmental management for vector control is one of the new strategies developed to tackle the menace of vectors. Manipulation of abiotic factors has gained acceptance due to laboratory and semi-field trial findings. In this chapter, the authors review the literature on the influence of some critical abiotic factors affecting the bionomics of immature and adult mosquito species. They also review prospects for developing protocols based on these findings.

I appreciate the great efforts of IntechOpen's Author Service Manager Ms. Maja Bozicevic throughout the development and publication of this book. This book is a useful resource for entomologist, parasitologists, researchers, scientists, and students, as well as growers, producers, and others that face the challenges imposed by Dipteran insects.

Dr. Farzana Khan Perveen (Gold medalist and FLS)
The President,
Classes et Events in Sciences (C.E.S.),
Avignon, France

Section 1

Introduction to Diptera

Chapter 1

Introductory Chapter: Diptera

Farzana Perveen and Anzela Khan

1. Introduction

Diptera is an order of single pair winged insects commonly known as flies/ true flies. The 2nd pair of wings are modified into halteres. They are mostly small to medium-size. Biologically, it is very vast order with greatly diversified insects. Many have co-evolved in association with plants and animals and most successful groups of organisms on this universe (**Figure 1**; **Table 1**) [2].

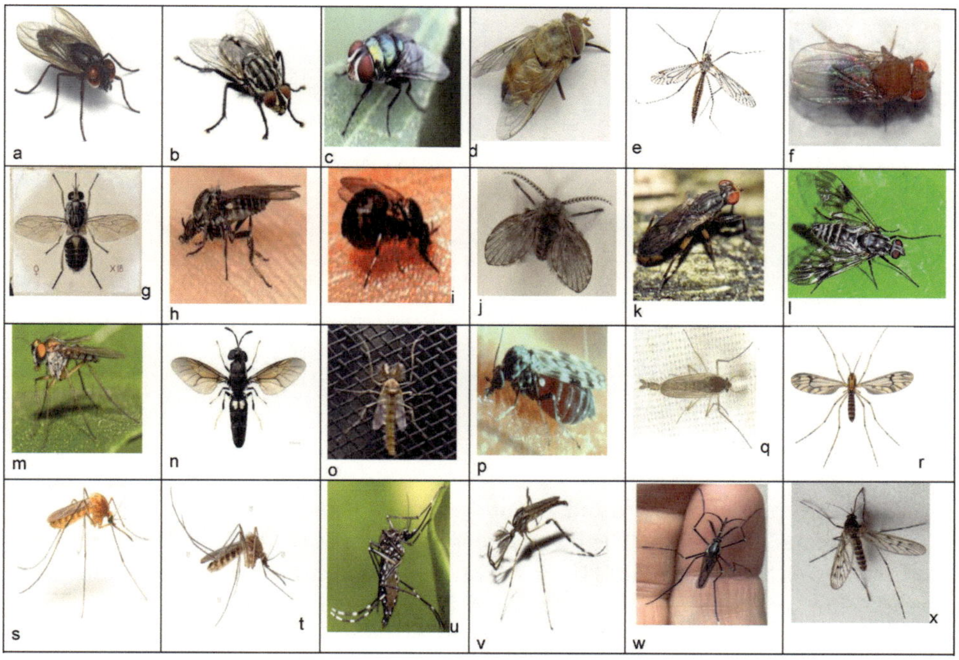

Figure 1.
Members of order Diptera: **Flies:** *a):* Housefly, Musca domestica *(Muscidae); b): Flesh fly,* Sarcophga carnaria *(Sarcophagidae); c): Blowfly,* Chrysomya megacephala *(Calliphoridae); d): Horse-fly,* Tabanus bovinus *(Tabanidae); e): Crane-flies,* Tipula oleracea *(Tiulidae); f): Fruit-fly,* D. melanogaster *(Drosophilidae); g): Tsetse-fly,* Glossina gambiensis *(Glossinidae);* **tinny-flies:** *h): Sand-fly,* Austrosimulium australense *(Simuliidae); i): Black-fly,* Parasimulium furcatum *(Simuliidae); j): Moth-fly,* Clogmia albipunctata *(Psychodidae); k): Marsh-flies,* Pherbellia annulipes *(Sciomyzidae); l): Watersnipe-fly,* Ibisia marginata *(Athericidae); m): Aquatic long-legged-fly,* Chrysosoma adoptatum *(Dolichopodidae); n): Soldier-fly,* Hermetia illucens *(Stratiomyidae);* **midges:** *o): Non-biting midge,* Chironomus *(Chironomidae); p): Biting midge,* Culicoides sonorensis *(Ceratopogonidae); q): Phantom-midge,* C. punctipennis *(Chaoboridae); r): Dixid midge,* Dixa nebulosa *(Dixidae);* **mosquitoes:** *Mucidae): s): African malaria mosquito,* A. gambiae; *t): Common-house-mosquito,* C. pipiens; *u): Yellow-fever-mosquito,* A. aegypti; *v): Shaggy-legged-gallinipper,* Psorophora ciliate; *w): Elephant mosquito,* Toxorhynchites rutilus; *x): Banded-house-mosquito,* C. annulata *[1].*

IntechOpen

Domains: **Eukaryota**
Super-kingdom: **Opisthokonta**
Kingdom: **Animalia** Linnaeus, 1758
Sub-kingdom: **Invertebrata**
Clade 1: **Metazoa** Haeckel, 1874
Super-Division: **Eumetazoa**
Division: **Bilateria**
Sister-clade: **Protostomia**
Sub-division: **Ecdysozoa**
Super-phylum: **Tactopoda**
Group: **Panarthropoda**
Phylum: **Arthropoda** Von Siebold, 1848
Clade 2: **Mandibulata**
Sub-phylum: **Atelocerata**
Super-class: **Hexapoda**
Class: **Insecta**
Infra-class: **Neoptera**
Sub-class: **Pterygota**
Unranked: **Endopterygota**
Ranked: **Holometabolous**
Super-order: **Panorpida**
Order: **Diptera** Linnaeus, 1758

Table 1.
Taxonomic position of order: Diptera [1].

2. History of Diptera

First true dipterans were found in middle Triassic, and widely spread during middle and late Triassic [3]. The basal clades in Diptera include Deuterophlebiid and mysterious Nymphomyiid [1]. Based on fossil record, 3 episodes of evolutionary radiation are thought to have happened. Numerous novel kinds of subordinate Diptera established in the Triassic, nearby 220 million years back. Several inferior Brachycera seemed about 180 million ages back in Jurassic. A 3rd radiation acquired mid Schizophora at commence of Paleogene, 66 million years past [4].

3. Classification

About 150,000 species are described in 150 families (**Figure 1**; **Table 1**) [2].

4. Morphology

4.1 Head segment

Head is distinct from thorax, with a marked narrowing at neck. The suture separates 2 regions, upper one is the frontal region, which has continuity with apex, orbital region and gena; lower one, the clypeus, contains the insertion of the antennae and ends with epitomal edge, which comprises the upper lip [3]. They have filiform, stylate or aristate antennae. All fly antennae consist of 3 parts: scape, pedicel, and flagellum [2]. They have prominent compound eyes on a mobile head. They grow to occupy most of the side of the head. The morphology of the compound eye is characterized by a significant number of ommatidia. The ocelli, when present are located in the top of the head, arranged at the corners of a triangle in an area called stemmaticum or ocellar triangle [2].

4.2 Mouthparts

They are adapted and joint into a sucking proboscis, which is extremely different in construction. The inherited state was the piercing and sucking sort of proboscis. There are more improved proboscis arrangements multifariously rasp or sponge fluids. The labellum is modified into sponge structure. Non-functional adult mouthparts are also found in several Diptera. In some species the mouthparts of the females are adapted for piercing the skin of hosts and feed on blood as ectoparasites [2].

4.3 Thoracic characteristics

Thorax is 2nd morphological region of the body in Diptera and it bears the locomotion organs, represented by 3 pairs of legs. In all Insects, it is composed by morphological and structural organization of the first 3 post-cephalic segments, named pro-, meso-, and metathorax, in order antero-posterior [2].

4.4 Wings: flying organs

They have one pair of functional and membranous wings are attached to the complex mesothorax [2]. Flies are capable of great maneuverability during flight due to the presence of 2nd pair of wings on the metathorax, are reduced to halteres. They provide fast response to the wing-steering tissues, therefore, they function as a balancing and controlling organs for the body. Flies are without of the halteres are not able to fly. The wings and halteres move simultaneously but the strength of individual wing strike is self-regulating, permitting the fly to try slanted [5]. The wings of the fly are connected with 2 types of fibrous tissues (FT), one is used to supply them with mechanical or electrical energy and second is used for well regulator [6]. Flies have an ability to fly in an upright direction, however, they rapidly diverge in different straight-line paths. Particularly, the change of directions encompasses a slant of 90° and accomplish within 50 milliseconds, it is known as saccades. They originate by optical impulse as the fly perceives an item, nerves then activate FT in the thorax that responsible a minor change in wing stroke, which produce enough rotation. It can be detected within 4–5 wing-beats; therefore, the halteres trigger a counter-turn and the fly heads off in a new direction [7].

4.5 Abdominal characteristics

The abdomen is 3rd part of the body of Diptera. It composed of 11 abdominal segments, called urites, the newest of which are compact and distinguished with sexual structures. A sole urite seems as a circle with dorsal sclerite, named as tergite or tergum. However, a ventral one, termed as sternite or sternum. They are connected by a pleural membrane. Each urite is connected to adjoin by an inter-segmental membrane. Morphology of the abdomen is substantially determined by morpho-anatomic adaptation, in both sexes, as a function of the reproduction. In feminine, the primordial urites grow delicately and elastically, they form a capricious enlarging ovipositor. This structural transformation frequently follows by sclerotization of 8th (last) urite, therefore, the ovipositor enters through the tissues of the host-organism, to house the eggs and larvae. In masculine, the same passes a complicated alteration to make an organ, united with genitalia titled the hypopygium [2].

4.6 Locomotion

Diptera have 3 pairs of legs, one pair on each of 3 segments of the thorax and are generally called the fore, mid-, and hind-legs. Diptera larva is apodous (without legs), but sometimes, especially in aquatic larva has appendages similar to pseudopodia. If fly is walking on the wall or ceiling, then it was observed that other portions of the tarsi in action. The bottom of the housefly's feet boasts tiny, gripping claws and moist suction pads called pulvilli, which allow the fly to land almost anywhere [3].

4.7 Development

They have complete metamorphosis (holometabolous) [3] (for detail concern Chapter: Characteristics of Dipteran Insects).

4.8 Habits

Food habits of some species are unknown but most of Diptera may feed wide varieties of materials. They are detritivores (common fly, *Dryomyza anilis* and housefly, *Musca domestica*), flower feeders [Acalyptratae, Bibionidae, Conopidae] and nectar feeders [Nemestrinidae, Bombyliidae and Tabanidae]. Adults Brachycera feed on flowers, Syrphidae, which obtain all their protein requirements by feeding on pollen. Both male and female mosquitoes feed on nectar and plant juices, but in many females suck the blood of mammals, birds, amphibians or reptiles to obtain proteins as materials for maturation of eggs and oviposition. Many Diptera are obligatory blood-feeders [Muscidae, Phlebotominae, and Rhagionidae] [2].

Their larvae feed on diverse nutrients, different from those of adults. They feed on leaf-litters, leaves, stems, roots, flowers and seed heads, mosses, fungi, rotting woods, fruits; other organic matters such as slime, flowing sap, rotting cacti, carrions, dungs, detritus in mammals, birds or wasp nests; fine organic materials including insect frasses and micro-organisms. Tachinidae larvae parasitise on insects. Endoparasites larvae of Conopidae feed on bees, wasps, cockroaches and calyptrates, however, Pyrgotidae feed on adult scarab beetles. Sciomyzidae larvae are exclusively associated with freshwater and terrestrial snails or slugs. Odiniidae larvae feed in the tunnels of wood-boring larvae of Coleoptera, Lepidoptera, and other Diptera. *Oedoparena* [a small genus of Dryomyzidae] feed on barnacles [2].

4.9 Habitats

Diptera occur all over the world except in regions with permanent ice-cover. They are abundant throughout the world and occupy virtually every terrestrial niche, in tropics, subarctic, at sea level, and high on mountains. They colonize on beaches to low-tide level, they also found into deeper water, and only 1–2 midges are truly marine. *Pontomyianatans* are found in the Pacific as well as in freshwater. On the other hand, migrating flies have been found far out to sea. They are found in most land-biomes, including caves, deserts, and tundra. Palearctic habitats

Figure 2.
Camouflage in Diptera: a): Hover-fly, Syritta pipienns *(Braulidae); b): Bee-louse,* B. coeca *(Braulidae); c): Soldier-fly,* Lasiopa villosa *(Stratiomyidae); mimicry in Diptera: d): European-hover-fly,* Pocota personata *(Syrphidae) with a bumble-bee (Hymenoptera); e): Gepunktete-hornfliege,* Trypetoptera punctulata *(Tephritidae) with spider (arachnids: Araneae); f): Members of Sepsidae with ant (Formicidae); g): Members of Stratiomyidae with wasp (Hymenoptera); h): Micropezid (Micropezidae) with green-tree-ant,* Oecophylla smaragdina*; i): Large-robust-flies,* Pantophthalmus rothschildi *(Pantophthalmidae) and giant-fly,* Mydas praegrandis *(Mydidae) with members of Pompilidae (Hymenoptera) [5].*

include meadows, prairies, mountain passes, forests, desert oases, seashores, sandy beaches, coastal lagoons, lakes, streams, rivers, bogs, and fens. They also found in areas polluted-water by rotting waste, industrial emissions, urban areas, cattle, horse and poultry farms [2].

4.10 Association

Stylogastrinae are obligatory associated with Formicides and Orthoptera and other Diptera. They typically use ground-dwelling Orthoptera and army ants (Formicidae) for raiding columns to flush out their prey. Phoridae are specialists parasitoids on ants, but tropical species are parasitoids on stingless bees (Hymenoptera). They are often host to more than one or 12 fly-larvae. More than 400 species of hoverflies (Syrphidae: Microdontinae) are myrmecophiles, they live in the nests of ants as scavengers or predators. Adults of many *Bengalia* (Calliphoridae: Bengaliinae) are kleptoparasites on ants and snatch foods and pupae being carried by ants or feed on winged termites [4].

4.11 Camouflage or mimicry

Camouflage is common in Diptera. Hover-flies, *Syritta pipienns* (Braulidae) use motion camouflage to approach females. Bee-louse, *B. coeca* (Diptera) uses chemical camouflage to survive within honeybee colonies. Soldier-fly, *Lasiopa villosa* (Diptera, Stratiomyidae) male shows camouflaged by hiding on a dry elm twig (**Figure 2a–c**) [6].

Many Diptera exhibit batesian mimicry. Syrphidae often are brightly colored, with spots, stripes, and bands of yellow or brown covering their bodies. Due to this coloring, and sometimes behavior patterns, they are often mistaken for wasps or bees (Hymenoptera). Wing pattern of *Trypetoptera punctulata* (Sciomyzidae) is very similar to some Tephritidae, therefore, they mimic the color pattern of some spiders (Arachnidae) [5]. Several fly species are look like an ant. Sepsisoma spp. (Richardiidae) mimic ants particularly with formicinae, *C. crassus*. Species of stilt-legged flies especially the wingless and haltere-less, *Badisis ambulans* (Micropezidae) resemble ants, as do species in *Strongylophthalmyia* and *Syringogaster spp*. Mydidae are mimics of stinging Hymenoptera (**Figure 2d–i**) [6].

5. Reproductive performance

Fly gives optical signals as diverse from biochemical or other signs throughout their sexual life.

5.1 Sexual-selection and courtship

Diptera show sexual-selection and numerous designs of sexual-dimorphism, such as stretching of male-body, eye-stalks, or addition of exoskeleton have changed frequently in flies. Fruit fly, *Phytalmia mouldsi* McAlpine and Schneider (Tephritidae) uses a resource defense mating-system. Mountain midges, *Deuterophlebia* Edwards males have extremely long antennae, which they employ when contesting territories over running water, waiting for females to mate. Acalyptratae (sub-section of Diptera) exhibit morphological development associated with agonistic behavior include: Clusiidae, Diopsidae, Drosophilidae, Platystomatidae, Tephritidae, and Ulidiidae [8].

5.2 Swarms

Swarm-based mating systems typically involve males flying in swarms to attract patrolling females. Such swarms are often of gigantic size. Smaller swarms may be around a fixed point called a swarm marker. Swarming occurs in Chronomidae, Bibionidae, Platypezidae, Limoniidae, Fanniidae, Chloropidae, Coelopidae, Milichiidae, and Trichoceridae. Chaoboridae form larval as well as adult swarms [8].

5.3 Bioluminescence

Keroplatidae and glow-worm, *Orfelia fultoni* (Mycetophilidae) display bioluminescence. In some, it is restricted to immature stages, but in others, this character is kept by pupae and adults. Ability to produce their own-light is used by some predatory-larvae as a bait for potential-prey, but it makes them more vulnerable to predation or parasitism [9].

6. Economic importance

Dipterans are an important group of insects with economic importance and have a considerable impact on the environment [10].

6.1 Useful Diptera

They are also use as bioindicator for environment, and biological control agents, some Diptera produce useful products, they are foods for other animals, they participate as pollinators, dispersal of seeds, and symbionts. They are extraordinary important insects in the putrefaction and deterioration of materials of plant and animal. They remain instrument for the disintegration and permit the nutrients mix into the humus to increase fertility of soil. The immature stages are supplementary diet for higher agrarian organisms. In food chains, they are also a significant constituent. Many larvae function as predators, parasitoids, or scavengers of the host larvae. The larvae of Acroceridae and some Bombyliidae are hypermetamorphic. Diptera are very advantageous to men. Houseflies, blowflies and fungus gnats (Mycetophilidae) are scavengers and assistance in decay. Robber (Asilidae and Tachinidae), dagger and balloon flies (Empididae) are predators and parasitoids. They control a diversified pest. Bee-flies (Bombyliidae) and hoverflies (Syrphidae) are pollinators of crops [11]. The following are special uses of Diptera:

1. Fruit-fly, *D. melanogaster* has long been used as a model organism in genetical researches, because it is easy to bred and reared in the laboratory (**Figure 3a**) [13].

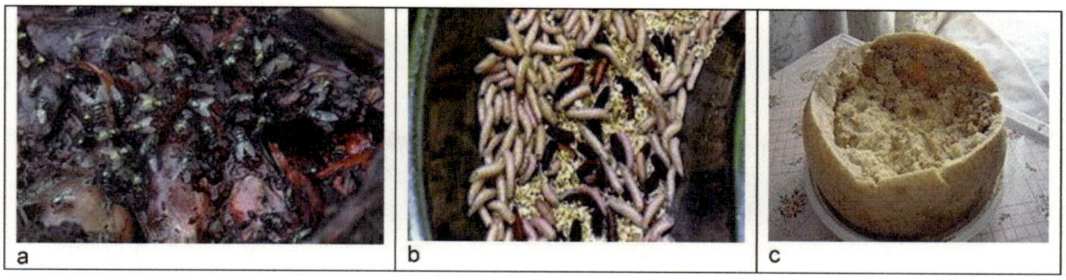

Figure 3.
Useful Diptera: a): Techniques and on molecular sequences in phylogenetics; b): Maggots are useful to forensic entomology; c): Sardinian cheese, casu marzu is exposed to cheese skippers such as Piophila casei *(Piophilidae); [12].*

2. They are also use for investigations in physiology, microbial pathogenesis and development among other research topics [14].

3. Maggots (Diptera larvae) can be used as a biomedical-tool for chronic wound-care as they are safe and effective. Eradicating dead tissues promote cell-growth and healthy-wound-healing. They have biochemical properties such as antibacterial activity found in their secretions [15].

4. Studies on dipteran by Willi Hennig used in the development of cladistics techniques that he applied to morphological characters, but now adapted for use with molecular sequences in phylogenetics [16].

5. Maggots found on corpses are useful to forensic entomology. They visit corpses and carcases at fairly well-defined times after the death of victim (**Figure 3b**) [17].

6. Maggots used as animal feed at zoological gardens and safari parks. Blow-fly larvae (gentles) and bluebottle larvae (casters) are produced commercially. They use as bait for fish, and as food for carnivorous, kept as pets, in zoos, or for research. They also use as large-scale food for farmed chickens, pigs, and fish [18].

7. Sardinian cheese, casu marzu is exposed to flies cheese skippers such as *Piophila casei* (Piophilidae). Digestive activities of the fly larvae soften the cheese and change the odor as fragment of course of ripening (**Figure 3c**) [19].

6.2 Harmful Diptera

The applied significance of the Diptera is as disease vectors, and agricultural pests. Many Diptera larvae are predatory. Some Tephritidae are leaf miners or gall formers. They are obligate parasites of mammals (Oestridae). There are roughly 150 known species worldwide those cause myiasis in animals. They are members of related families, such as the Calliphoridae [20].

6.3 Diptera as pests

Some leaf-miner flies (Agromyzidae), fruit flies (Tephritidae and Drosophilidae) and gall midges (Cecidomyiidae) are pests of agricultural crops; others such as tsetse flies, screwworm and botflies (Oestridae) attack livestock, causing wounds, spreading disease, and creating significant economic harm [20].

6.4 Diptera as medical importance

Mosquitoes (Culicidae), black-flies (Simuliidae) and drain-flies (Psychodidae) have great impact on human health as vectors of major tropical diseases. *Anopheles* mosquitoes transmit malaria, filariasis, and arboviruses; *A. aegypti* mosquitoes carry dengue-fever and Zika-virus; black-flies carry river blindness; sand-flies carry leishmaniasis. Other dipterans irritate to humans, when present in large numbers. These include house-flies, which pollute foods and feast food-borne diseases such as biting midges and sand-flies (Ceratopogonidae) and the house-flies and stable-flies (Muscidae). In tropical regions, eye-flies (Chloropidae), which visit the eye in search of tears irritate in some seasons [20].

6.5 Emblematic Diptera

In different cultures, flies play a variety of symbolic roles. They have both good and bad impacts on faith. In the traditional Navajo religion, big-fly is an important spirit-being. In Christian, demonology, Beelzebub is a demonic-fly, Lord of the flies, and God of the Philistines. They have appeared in literature since ancient Sumer [21]. In 1962, the biologist Vincent Dethier describes the characteristics of flies in his book titled to know a fly. Design of miniature-flying-robots has been made due to inspiration from flies [22]. In 1993, Steven Spielberg's film Jurassic Park trusted on the idea that DNA could be preserved in the stomach contents of a blood-sucking-fly fossilized in amber, though the mechanism has been discounted by scientists [23].

7. Communication

Sex pheromones are known for many dipteran species and play an important role in courtship behavior, together with visual, tactile, acoustic and other factors. Pheromones for a number of dipterans have been recently identified. In the Nematocera pheromones are volatile components, which act at a distance [24].

7.1 Parthenogenesis and viviparity

Hippoboscidae (louse flies, sheep keds), Streblidae and Nycteribiidae (together known as bat flies) and tsetse flies (*Glossina* spp.: vectors of African trypanosomes) are distinguished by their specialized reproductive biology, defined by adenotrophic viviparity (maternal nourishment of progeny by glandular secretions followed by live birth) is unique reproductive mechanism [25]. A study of chromosomal variability in the diploid parthenogenetic species *Lonchoptera dubia* (Lonchopteridae, Brachycera, Diptera) is based on an analysis of 272 females from 32 widely separated areas, chiefly in eastern North America. *L. dubia* is almost entirely nearctic and northern in distribution, occurring in the northern United States and southern Canada. Selection studies on the automictic (normally bisexual) tychoparthenogenetic species *Drosophila parthenogenetica* indicate that continued selection for parthenogenesis within a unisexual line for 62 generations results in selective improvement in parthenogenesis during the first 17 generations, and thereafter no change in rate of parthenogenesis [26].

7.2 Hibernation and diapause

Eastern tree hole mosquito, *A. triseriatus* 4th-instar larvae pupate only under constant increased daylength (longer day in photoperiodism). In common house mosquito, *C. pipiens*, females normally digest blood meal (trypsin and a chymotrypsin pathway), as the females then enter diapause. In *C. pipiens*, there are about 40 genes which are upregulated and downregulated during diapause, these genes code for functions like regulatory functions, metabolic functions, digestion, endocrine functions, cytoskeletal genes, ribosomal genes, transposable elements, and other with unknown functions. Asian tiger mosquito, *A. albopictus* collection was least during winter in Hanoi, Northern Vietnam, as it undergoes diapause. Hibernation mostly occurs in insects living in polar regions. *Parous Cx. p. pipiens* females from region of the northeastern US enter hibernacula during winter [27].

8. Internal anatomy and physiology

8.1 Nervous system

The nervous system of Diptera can be divided into a brain and a ventral nerve cord. The head capsule is made up of 6 fused segments. The first 3 pairs of ganglia are fused into the brain, while the three following pairs are fused into a structure of 3 pairs of ganglia under the insect's esophagus, called the subesophageal ganglion. *Musca domestica*, have all the body ganglia fused into a single large thoracic ganglion. At least a few insects have nociceptors, cells that detect and transmit sensations of pain (**Figure 4**) [29].

8.2 Digestive system

An insect uses its digestive system for all steps in food processing: digestion, absorption, and feces delivery and elimination. Main structure of Diptera's digestive system is a long enclosed tube called alimentary canal (or gut), which runs lengthwise through the body and directs food in one direction: from the mouth to the anus. It can be divided into 3 sections: the foregut, midgut and hindgut, each of which performs a different process of digestion (**Figure 4**) [29].

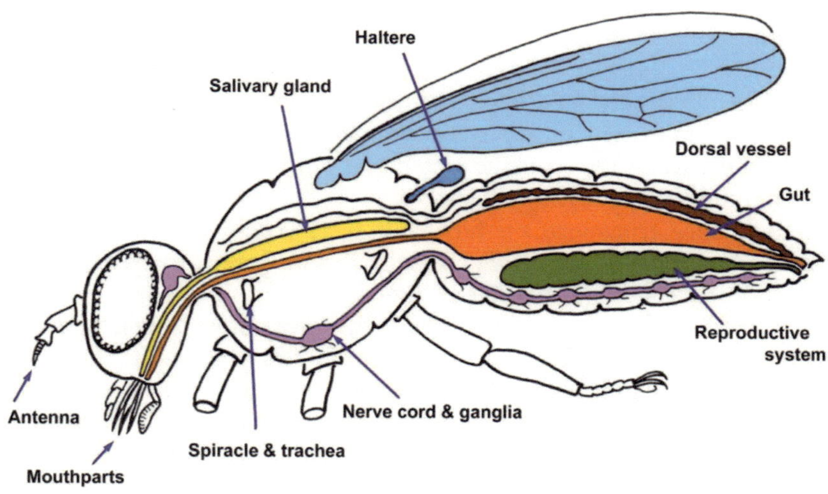

Figure 4.
Morphology and physiology of Diptera: Nervous system, digestive system, excretory system, respiratory systems, circulatory system, endocrine system, and reproductive system [28].

8.3 Excretory system

In the hindgut, undigested food particles are joined by uric acid to form fecal pellets. The rectum absorbs 90% of the water in these fecal pellets, and dry pellet is then eliminated through anus. Uric acid is formed from haemolymph waste products diffused from the Malpighian tubules at the junction between mid- and hindgut ranging from only 2 to over 100 tubules (**Figure 4**) [29].

8.4 Respiratory systems

Diptera respiratory system composed of internal tubes and sacs through which gases either diffuse or are actively pumped directly to tissues that are trachea. Since oxygen is delivered directly, the circulatory system is not used to carry oxygen (**Figure 4**) [29].

8.5 Circulatory system

A single, perforated dorsal tube that pulses peristaltically, toward the thorax, the dorsal tube divides into chambers and acts like the insect's heart. The opposite end of the dorsal tube is like the aorta circulating the haemolymph, inside the body cavity. Air is taken in through openings on the sides of the abdomen called spiracles. Haemolymph's main function is that of transport and it bathes the insect's body organs. It transports hormones, nutrients and wastes and has a role in, osmoregulation, temperature control, immunity, storage (water, carbohydrates and fats) and skeletal function. It also plays an essential part in the molting process. Body fluids enter through one-way valved ostia, which are openings situated along the length of the combined aorta and heart organ. Movement of haemolymph is particularly important for thermoregulation in Diptera (**Figure 4**) [29].

8.6 Endocrine system

Endocrine system of Diptera composed of neurosecretory cells, corpora cardiaca, prothoracic glands, and corpora allata (**Figure 4**) [29].

8.7 Reproductive system

Female insects are able make eggs. The ovaries are made up of a number of egg tubes, called ovarioles, which vary in size and number by species. The number of eggs that the insect is able to make vary by the number of ovarioles. Accessory glands or glandular parts of the oviducts produce a variety of substances for sperm maintenance, transport, and fertilization, as well as for protection of eggs. Spermathecae are tubes or sacs in which sperm can be stored between the time of mating and the time an egg is fertilized (**Figure 4**) [29].

Main male reproductive organ is the testis. However, most male Diptera have a pair of testes, inside of which are sperm tubes or follicles that are enclosed within a membranous sac. The follicles connect to the vas deferens by the vas efferens, and 2 tubular vasa deferentia connect to a median ejaculatory duct that open outside. A portion of the vas deferens is often enlarged to form the seminal vesicle, which stores the sperm before they are discharged into the female. The ejaculatory duct is derived from an invagination of the epidermal cells. The terminal portion of it may be sclerotized to form the intromittent organ, the aedeagus. The aedeagus can be quite pronounced or de minimis (**Figure 4**) [29].

Author details

Farzana Perveen[1*] and Anzela Khan[2]

1 Classes et Événment en Sciences (C.E.S.), Avignon, France

2 University of Wollongong (UOW), KDU, Malaysia

*Address all correspondence to: farzana_san@hotmail.com

References

[1] Downes WL Jr, Dahlem, GA. Keys to the Evolution of Diptera: Role of Homoptera. Environmental Entomology. 1987;16 (4):847-854.

[2] Colless DH, McAlpine DK. *Diptera* (flies). In: The Division of Entomology. Commonwealth Scientific and Industrial Research Organisation, Canberra (spons.), The insects of Australia. Melbourne Univ. Press, Melbourne. 1991;717-786.

[3] Blagoderov VA, Lukashevich ED, Mostovski MB. Order Diptera Linné, 1758. The true flies. In Rasnitsyn, A. P.; Quicke, D. L. J. (eds.). History of Insects. Kluwer Academic Publishers. 2002;1-66.

[4] Wiegmann BM, Trautwein MD, Winkler IS, Barr NB, Kim JW, Lambkin C, Bertone MA, Cassel BK. Episodic radiations in the fly tree of life. PNAS. 2011;108 (14):5690-5695.

[5] Deora T, Singh AK, Sane SP. Biomechanical basis of wing and haltere coordination in flies. Proceedings of the National Academy of Sciences. 2015;112 (5):1481-1486.

[6] Dickinson MH, Tu MS. The function of dipteran flight muscle. Comparative Biochemistry and Physiology: Part A: Physiology. 1997;116 (3):223-238.

[7] Dickinson, MH. The initiation and control of rapid flight maneuvers in fruit flies. Integrative and Comparative Biology. 2005;45 (2):274-281.

[8] Yeates DK, Wiegmann BM. The Evolutionary Biology of Flies. Columbia University Press, Columbia, USA. 2005; 34-90.

[9] Perveen F, Khan A. Characterization of the insect-fauna of the free-ranging urban dog, *Canis domesticus* (L.) carcass in the tropical region of Pakistan: A tool for forensic entomology. Advances in Entomology. 2013;1 (2): 29-37.

[10] Perveen F, Khan A. Ecological succession pattern of the insect species associated with the carcass of the free-ranging urban dog, *Canis domesticus* (L.): A tool for forensic entomology. Mintage Journal of Pharmaceutical and Medical Sciences. 2013;2 (3):31-36.

[11] Ali PA, Zahid M, Shah M, Sthanadar AA, Ahmad A, Mehmood T, Perveen F, Shah M. (2013) Forensically important Diptera species associated with dog carcass (*Canis domesticus* L.) for a case study in District Mardan, Pakistan. International Journal of Biosciences. 2013;3 (6):128-134.

[12] Resh VH, Cardé RT. Encyclopedia of Insects. Academic Press. 2009;284-297.

[13] Perveen FK. (Ed.) *Drosophila melanogaster*: Model for Recent Advances in Genetics and Therapeutics. InTech Book Publisher, Rijeka, Croatia. 2018; 1-257.

[14] Perveen F, Asmat P. Decomposition of the free-ranging urban dog, *Canis domesticus* (L.) carcass with reference to insect fauna identification: A tool for legal investigations. Malaysian Journal of Forensic Sciences. 2014;5 (1):69-77.

[15] Perveen F, Ali PA, Akbar MF. Key for identification of insect-fauna of the free-ranging urban dog, Canis domesticus (L.) carcass in tropical region of Pakistan: A tool for forensic entomology. Pakistan Journal of Entomology, Karachi. 2013;28 (2):117-126.

[16] Mulla MS, Chansang U. Pestiferous nature, resting sites, aggregation, and host-seeking behavior of the eye fly, *Siphunculina funicola* (Diptera: Chloropidae) in Thailand. Journal of Vector Ecology. 2007;32 (2):292-301.

[17] Online. Why use the fly in research? YourGenome. 2015;1-6. (Retrieved: 27 May 2021).

[18] Ashlock PD. The Uses of Cladistics. Annual Review of Ecology and Systematics. 1974;5 (1):81-99.

[19] Joseph I, Mathew DG, Sathyan P, Vargheese G. The use of insects in forensic investigations: An overview on the scope of forensic entomology. Journal of Forensic Dental Sciences. 2011;3 (2):89-91.

[20] Blagoderov VA, Lukashevich ED, Mostovski MB. Order Diptera Linné, 1758. The true flies. Ed. Rasnitsyn AP, Quicke DLJ. History of Insects, Kluwer Academic Publishers. 2002; 234-567.

[21] Arndt W, Danker FW, Bauer W. A Greek-English lexicon of the New Testament and other early Christian literature (3rd ed.) University of Chicago Press, Chicago, USA. 2000;5 (173):1-56.

[22] Insect-scale robot. Science. 2013;340 (6132):603-607.

[23] Gray R. Jurassic Park ruled out-dinosaur DNA could not survive in amber. Daily Telegraph. 2013;3-4.

[24] Thomas CW. Pheromonal communication involved in courtship behavior in Diptera. Journal of Insect Physiology. 2007;53 (11):100-1089.

[25] Joshua BB, Geoffrey MA, Aaron AB, Veronika M, Serap A. Adenotrophic viviparity in tsetse flies: Potential for population control and as an insect model for lactation. Annual Review of Entomology. 2015;7 (60):351-371.

[26] Harrison DS. On the Evolution of Parthenogenesis in Lonchoptera (Diptera). Evolution. 1956;4 (10): 345-359.

[27] Chapman RF. The insect structure and function, Edition 3rd, ELBS publisher. Clements. A.N. Physiology of Mosquitoes. 1982;19 (63):220-232.

[28] Online. https://www.welcomewildlife.com/all-about-flies. (Retrieved: 2 June 2021).

[29] Hennig W. Flügelgeäder und system der Dipteren. Beiträge zur Entomologie. 1954;4: 2453-88.

Chapter 2

Characteristics of Dipteran Insects

Murat Helvacı

Abstract

Diptera means two wings (Di: two, pteron: wing). They have complete metamorphosis and they are holometabolous insects which means there are 4 stages (egg, larvae, pupae and adult). The name of larval stage is "maggot". Some of the dipteran insects cause damage in agricultural production. Some are harmful for humans. Dipteran insects have two wings. Hind wings are reduced and they are called "halteres". Function of halteres is balancing when the insects fly. Except mosquitoes, dipteran insects have sponging-sucking mouthparts. Important examples for dipteran insects are Olive fruit fly and Medfly which cause damages in agricultural production. OFF is the most destructive pest in olive growing areas and Mediterranean fruit fly cause damages in fruit production.

Keywords: characteristics, diptera, haltere, holometabola, mouthpart

1. Introduction

Many insects are called flies such as butterflies and dragonflies, but only insects belonging to the order diptera are known as "true flies." Dipteran insects (flies and mosquitoes) are holometabolous insects which means they have complete metamorphosis life cycle. Dipteran insects have 4 stages in their life cycle (adult, pupae, larvae and egg). Name of larval stages of these insects is "maggot". Adults of this order are recognized according to wing types. Front wings are developed for flying. Other pairs of wings are undeveloped and they have balancing function when insects fly. These type wings give order its name: two (di-), wings (ptera). One pair of wing provides flying of insects and other pair developed into balancing stuructures. The name of other pair of wings which provide balancing during flight, called "halteres". Except mosquitoes, dipteran insects have sponging-sucking mouthparts. Mosquitoes and some other have piercing-sucking mouthparts. The Diptera order is divided into two or three subclasses: Nematocera and Brachycera, the second being Orthorrhapha and Cyclorrhapha [1]. There are approximately 152.000 identified species in this order which are distributed about 130 families [2]. Houseflies, hoverflies, mosquitoes and fruit flies are the most important species which belong to the diptera order. Houseflies carry diseases such as cholera and dysentery. The fly cause serious problems by carrying disease organisms onto food. They take disease organisms from leg hair or eat them and then feed them to other foods. The adults of hoverflies feed mainly on pollen grains. The larvae of some species eat rotten plant and animal materials in the soil or in ponds, lakes and streams.

Another species of hoverflies's larvae are predators and feed on some harmful insects such as aphids and thrips etc. So, the larvae of these species are important for biological control as natural enemies.

The females of mosquitoes suck the blood of human and several animals such as farm animals and wild animals. Mosquitoes are vector insects and cause the spread of serious illnesses such as malaria and leishmaniasis etc. [3]. Fruit flies are the most destructive pests in horticultural growing. For example, Olive fruit fly are the main pest in growing areas where table and olive oil production are made throughout the world.

General objectives of this chapter is to provide information about the habitats, feeding patterns and life cycles of insects belonging to the order Diptera, and the stages in which they exist in their life cycle; However, it is to provide information on the characteristics of each stage (adult, egg, larva and pupa) in the life cycle of these insects. In addition, to give information about the role of insects in this order for human and animal health and agricultural production.

2. Characteristics of Diptera insects

Dipteran order has 125.000 insect species and is one of the biggest order throughout the world and is highly diverse. Our world score is based on data from higher than 152.000 known species and higher than 130 identified families and these data are from the"World Diptera's Bio Systematic Database" [2]. Diptera order has the fourth place after the Lepidoptera, Hymenoptera and Coleoptera. Many species belonging to the order Diptera are found in almost all zoogeographic regions of the world. These species are well adapted to a broad range of habitats. Except the depth of oceans, they can live many habitats on earth [1]. Maximum nutrient and biomass formation occurs in the maggots of Dipteran order, and the adult this order usually receives the energy which they need to feed their muscles which have functions for flying. Widely the broad range of insects (flies) looking for food, the food of those insects includes honey extract or nectar, blood of vertebrate, pollen, hemolymph of other insects, and another liquefied or liquefied biological resources which are suspended or dissolved in vomiting fluid or saliva. Few groups of adults are predators. Other few groups are completely devoid of mouthparts, so they do not receive food. Therefore, they have short life span. Larval stage of many species, as they exist, live in water, alive plant tissues and rotten organic matter. Moreover, they live as parasites or parasitoids of several animals. These larvae need a humid and wet atmosphere. In addition, the eggs these species hatch in water surfaces and larval stage occurs on the water surfaces. Maximum larvae live freely, also can be found in water, sediments, trees, fruits or decaying biological material, while other larvae are found in the tissues of living beings [2].

3. Life cycle of Diptera insects

Diptera insects are holometabolous (complete metamorphosis) which means that they have 4 stages in their life cycles (adult, egg, larvae and pupae). Adult females lay their eggs and the amounts of eggs vary according to species. The females of some species lay a few eggs, some other lay thousands of them. Generally, females lay their eggs near the water and lay them as a group or singly.

3.1 Egg

The eggs are laid grouped or individually by adult females, and the females usually lay their eggs in water and are sometimes attached to materials. Except for diapause eggs, eggs tend to stand only last for a few days, which are used to prevent lack of water or unwanted temperatures in the ecosystem [4].

Mosquitoes which belong to the order of diptera and Culicidae family, lay their eggs in bunches or can lay singly in or near water (**Figure 1**). In contrast, some dipteran insects such as Olive fruit fly (*B. oleae* Rossi) (**Figure 2**), Medfly (*C. capitata* Wiedeman) which belong to the Tephritidae family and Spotted wing Drosophila (*D. suzukii* Matsumura) which belongs to Drosophilidae family, lay their eggs inside the developed fruits.

Usually, adult flies lay their eggs, which will pass into the larval stage within a few hours or days after hatching. Amounts of eggs which are laid by adult females vary between 1 and 250.However, multiple sequential egg batches can be made. Females of Medfly lay 300 eggs in her lifetime. Besides, the Green bottle fly (*Lucilia sericata* Meigen) lay about 2000 eggs in confinement. However, in a casual environment where energy and time are beginning to look for suitable areas to lay their eggs, the total number may be less than 1000 [2].

3.2 Larvae

Larvae of dipteran insects are easily known by the absence of thoracic structures (legs). This type of larvae is called Apod larvae. Body parts of larvae are usually fleshy (thorax and abdomen). The whole body of larvae is tubular and long. Larvae are approximately 2 Or 25 mm, but some species can be reached about 10 cm length. There are 12 segments in the bodies of larvae and that segmentation pattern is most common. 3 of them is found in thorax and 9 of them in the abdomen [4]. Some true midges, commonly seen in anoxic habitats such as blood worms, have an pigment which has invertebrate form and respiration function. This is called "hemoglobin" that helps to capture oxygen molecules [4] (**Figure 3**). After the eggs hatch, the larvae of the maximum species pass 3 to 4 stages on land or close to the bottom or above the water surface. For instance, females of Olive fruit fly And Spotted wing Drosophila lay their eggs inside the developing fruit [5] (**Figures 4** and **5**). After the emerging of larvae, they consume the rotten material in which they are laid. They eat more foods to store energy and nutrients for pupa stage [1]. Larval period is completed from nearly 2 weeks to several months. Larvae of Diptera insects do not have wing pads but are found in pupae [4]. Respiration process takes place above the skin of many larvae of dipteran insects. There are small gills above the skin of some

Figure 1.
Standing water mosquito and eggs.

Figure 2.
Egg of olive fruit fly inside olive fruit.

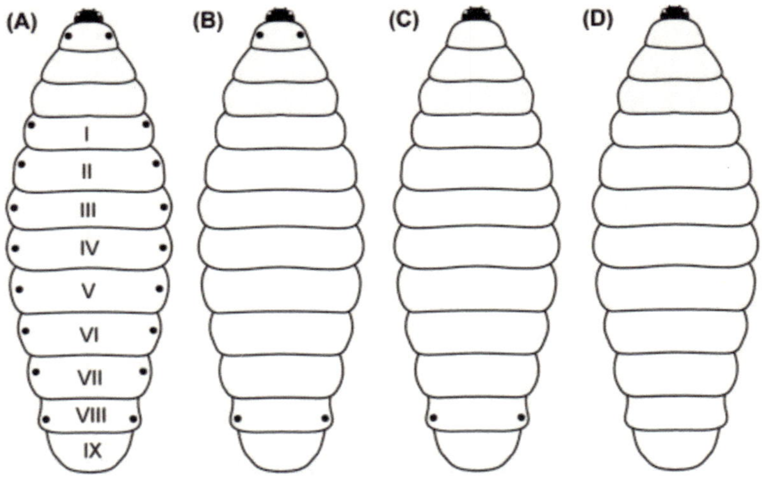

Figure 3.
Places of respiratory spikes in Diptera larva. (A) Peripneustic, (B) Amphipneustic, (C) Metapneustic, (D) Apneustic.

taxa. Other larvae of dipteran insects have spikes and they absorb oxygen from atmosphere by using long or short breathing tubes.

3.3 Pupae

Pupae of dipteran insects have a non-functional mandible (adecticeous). Their appendages can be independent from their body (externally) or attached to the body (obtect). Pupae of exarate types are hidden inside the hardened skin (puparium) of the final stage of larvae [2].

The pupae stage of dipteran insect varies significantly in shape. The pupae of some flies look like cross shape between larvae and adult, while other pupae shape of flies are featureless and they have a structure similar to seeds. The first forms are typical for Nematocera and are defined as having obtect or body-attached

Figure 4.
Larva of the olive fruit fly.

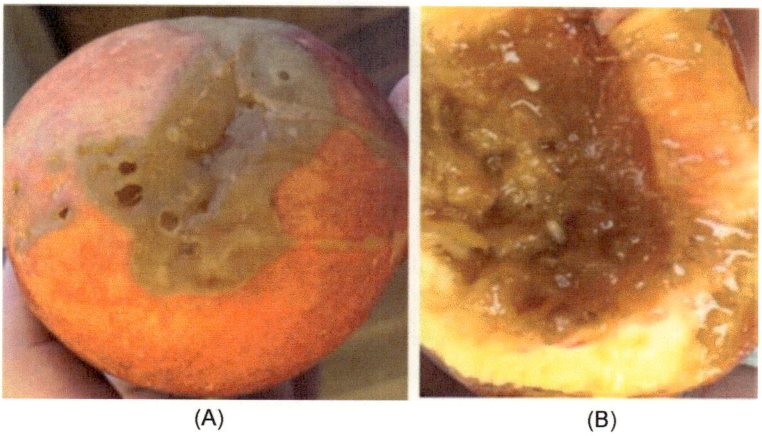

Figure 5.
Spotted wing drosophila: (A) egg-laying areas and (B) larvae of spotted wing drosophila.

appendages. For example, the pupae of a Crane fly (Tipulidae) have identifiable head, thorax and abdomen, but covers of antennae, legs and wing pads attach to the body of pupae. The exterior of the Nematocera pupa can be decorated with spines, breathing apparatus which are similar to gill or locomotory paddles.

Brachycera and Cyclorrhapha form the pupal stage in a different and more discreet way. The so-called higher Diptera family produces pupae that are described as coarctates, meaning "compressed" or "constricted". These taxa (eg Syrphidae, Drosophilidae, Muscidae) form a puparium consisting of hardened skin of the late larval stage [6] (**Figure 6**). The pupation of some flies occurs in the olive fruit or under the soil [5]. After the oviposition, the eggs hatch and the larvae feed on the fleshy part of the fruit, but leaves the fruit when ripe and some continue to appear inside the fruit. Larvae fall to the ground and pupation takes place in the soil [7].

3.4 Adult

Adults of dipteran insects have segmented body which includes head, thorax and abdomen parts. They have compound eyes which are found both side of head [1]. The size and shape of compound eyes are highly variable.

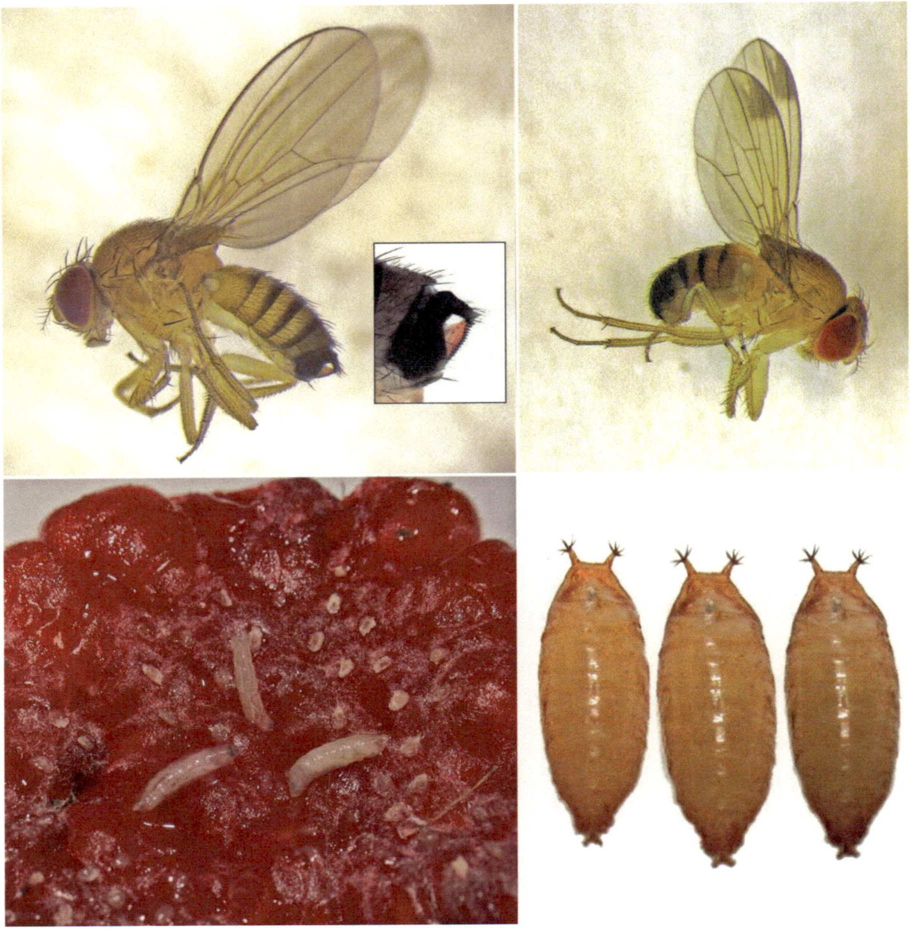

Figure 6.
Female, serrate ovipositor, male with typical black spots, larvae inside the fruit and pupae of Spotted wing *drosophila.*

Adults have dark reddish to black color compound eyes. Some families of dipteran insects have crossbands or spots of different colors such as Tabanidae, Syrphidae, Tephritidae, Sciomyzidae families [8].

There is small space in front of the head of adult dipterans. The function of this small space is to help the adults to see wider area when insects fly. Body color of adults changes from brown to black, orange or yellow, depending on the dipteran species [1]. For instance, body color of adults of *B. oleae* is brown and there are black spots in two sides of thorax and abdomen parts (**Figure 7**).

Thorax color of Spotted wing Drosophila is pale brown and there are horizontal black lines in the abdomen. Males have spots on their wings. Adult males of the *D. suzukii* are easily recognized a single black spots which are found on the outer edges of wings. Furthermore, two dark spots (sex comb) are found both of the forelimbs. Adult female of *D. suzukii* has a long, sclerotic, serrated ovipositor, unblemished open wings and comb on their feet [9] (**Figures 7** and **8**). Adult females of dipteran insects have ovipositor which is found at the end of abdomen. Females use this organ to lay their eggs inside the fruit. Females of OFF and Medfly lay their eggs (oviposition) inside the mature fruits by using their ovipositors.

Length range varies of adults varies according to dipteran species between 1 to 12 mm, but relative huge species are between 25–60 mm [4]. Sensory organ is found in front of the head of adults which is called antennae. Antennae are filiform

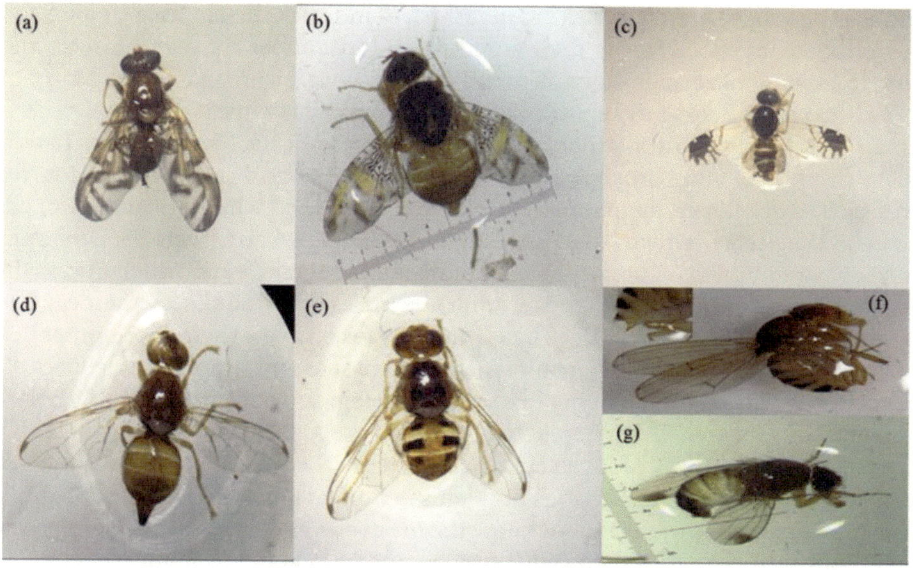

Figure 7.
General view of fruit fly adults. Euleia heraclei *(a),* Ceratitis capitata *(b),* Trupanea amoena *(c),* Bactrocera zonata *(d),* Bactrocera oleae *(e),* Drosophila suzukii *female and serrated egg laying organ (f) and male (g).*

Figure 8.
Spotted wing drosophila: (A) the black spots in the wings of male, (B) female, (C) egg laying organ (ovipositor) of female.

and front wings are developed to fly and hind wings are undeveloped (halteres) to balance when insects fly and tarsi 5-segmented [2].

4. Conclusion

Dipteran order has 125.000 insect species and is one of the biggest order throughout the world and is highly diverse. Diptera order has the fourth place after the Lepidoptera, Hymenoptera and Coleoptera. Houseflies, hoverflies, mosquitoes and fruit flies are the most important species which belong to the diptera order. Fruit flies cause destructive damages in agricultural production. Besides, houseflies transmit serious diseases transmits diseases by carrying disease organisms onto food such as cholera and dysentery. Mainly, adult hoverflies consume pollen grains. Larvae of some species feed on rotten plant and animal materials in lakes, streams and ponds or inside the soil. Another species of hoverflies's larvae are predators and feed on some harmful insects such as aphids and thrips etc. So, the larvae of these species are used in the biological control of harmful insects as natural enemies. The females of mosquitoes have piercing-sucking mouthparts and suck the blood of human, livestock and animals by using their mouthparts. Mosquitoes are vector insects and cause the spread of serious illnesses such as malaria and leishmaniasis etc. Many of the flies which belong to Tephritidae and Drosophilidae families, cause serious damages in the agricultural production such as Olive fruit fly, Medfly, Peach fruit fly, Celery fly Spotted wing drosophila. Dipteran insects can be recognized by some features such as developed membranous front wings and hind wings are undeveloped and called "halteres" which have functions as balancing when insects fly. Dipteran insects have complete metamorphosis (holometabolous) life cycle which means that there are 4 stages (egg, larvae, pupae, adult). Females of the adults lay their eggs into the food source or water. Eggs hatch and larvae complete their development. Pupation occurs under the soil, plant and animal tissues and water. The richness of the species in this order, the living in different ecological conditions and the morphological differences show people that this order is economically important and externalizes the diversity of invertebrate creatures in the world. There are insects in the Diptera order that cause serious problems in human, animal health and agricultural production. It is important to know the life cycles and habitats of these insects. Accordingly, the issues are followed (life cycle and habitats) to minimize disease transmission and damage in agricultural production. Another studies which will do in the future, will be useful in determining the economic effects of diptera insects on human health and agricultural production.

Conflict of interest

Author has no conflict of interest.

Author details

Murat Helvacı
European University of Lefke, Lefke, TRNC

*Address all correspondence to: mhelvaci@eul.edu.tr

References

[1] Badii BK. Phylogeny and Functional Morphology of Diptera (Flies). In: Sarwar M, editor. Handbook of Life Cycle and Development of Diptera. Open access peer-reviewed chapter; 2020. DOI: 10.5772/intechopen.90421.

[2] Sarwar M. Typical Flies: Natural History, Lifestyle and Diversity of Diptera. In: Sarwar, M, editor. Handbook of Life Cycle and Development of Diptera. Open access peer-reviewed chapter; 2020. DOI: 10.5772/intechopen.91391.

[3] Thorp JH, Rogers DC. Midges, Mosquitoes, Blackflies, and Other True Flies: Insect Order Diptera. Handbook of Field Guide to Freshwater Invertebrates of North America. Academic Press, Elsevier; 2011. p. 247-260. https://doi.org/10.1016/B978-0-12-381426-5.00027-2.

[4] Fusari L, Dantas GPS, Pinho LC. Order Diptera. Handbook of Thorp and Covich's Freshwater Invertebrates: Keys to Neotropical Hexapoda 4th ed. Elsevier; 2018. p.607-623. https://doi.org/10.1016/B978-0-12-804223-6.00029-9.

[5] Çatal BO, Helvacı M, Çalışkan Keçe AF, Ulusoy MR. First record of Drosophila suzukii in the areas of the Republic of Cyprus not under the effective control of the government of the Republic of Cyprus. Bulletin OEPP/ EPPO Bulletin, 2018; 0 (0), 1-3. DOI: 10.1111/epp.12464.

[6] Orsini MM, Daane KM, Sime KR, Nelson EH. Mortality of olive fruit fly pupae in California. Biocontrol Science and Technology, 2007; 17(8),797-807. DOI: 10.1080/09583150701527359.

[7] Resh VH, Carde RT, editors. Encyclopedia of Insects. 2nd ed. Academic Press is an imprint of Elsevier; 2009. p. 1169. ISBN: 978-0-12-374144-8.

[8] Merz B, Haenni JP. Morphology and terminology of adult Diptera (other than terminalia). In: Papp L, Darvas B, editors. Contribution to a Manual of Palaearctic Diptera (with special reference to flies of economic importance). Vol. 1.General and Applied Dipterology, Budapest: Science Herald.p.21-51.

[9] Byron MA, Gillett-Kaufman JL. Olive Fruit Fly Bactrocera oleae (Rossi) (Insecta: Diptera: Tephritidae) [Internet]. 2019. Available from : https://edis.ifas.ufl.edu/pdffiles/IN/IN27000.pdf.

Chapter 3

Fruit Flies (*Drosophila spp.*) Collection, Handling, and Maintenance: Field to Laboratory

Pragya Topal, Divita Garg and Rajendra S. Fartyal

Abstract

As drosophilids are versatile, low maintenance and non-harming model organisms, they can be easily used in all fields of life sciences like Genetics, Biotechnology, Cancer biology, Genomics, Reproductive biology, Developmental biology, Micro chemical studies, ecology and much more. For using such a model organism, we need to learn capturing, rearing and culturing their progeny along with basic identification and differentiation between males and females. This chapter is being emphasized on techniques of capturing these flies with different and effective techniques. Along with it, most species-specific baits are discussed to catch more yield. Culture food media, a set measurement of different ingredients is used to rear the collected sample. The reasons for using each ingredient are also discussed in this chapter. At last, this chapter highlights the basic clues to identify different species in the field and lab along with learning distinguishing characteristics of males and females easily and effectively.

Keywords: attractive baits, culturing flies, food preparation, identification, sorting out male and female

1. Introduction

In 1911 T.H. Morgan and his students C. Bridges, H. J. Muller and A. H. Sturtevant came across red-eyed insects, fruit-flies. Since then, it has made its way in research labs helping scientists to explore fundamental problems in biological sciences. It has turned out as one of the best metazoan insect model organisms. It is one of the best model organisms for the biological studies ranging from molecular genetics of diseases to the ecosystems and up to the evolutionary scales. Starting from visible mutants and chromosome mapping, today studies of complex genetic networks are possible with the help of multiple genome sequences (the 12 genomes project), systematic gene disruption or knock-down (RNAi stock library), microarray analysis, protein interaction maps and the FlyBase integrated database. High-throughput platform biology and open-source availability are considered as part of modern developments and the *Drosophila* model has integrated them with success. Recent developments allow the analysis of problems and processes previously inaccessible like complex human diseases caused by developmental, neurological or metabolic defects. *Drosophila* has clear-cut advantages in this field of research with its sophisticated genetic techniques. *Drosophila* research also plays an important role in technological transfer to other arthropod models, opening the window to biodiversity resources and macro-evolutionary scale. It has also been successfully integrated in teaching subjects as diverse as genetics, physiology, ecology and evolution.

Drosophilids are ectothermic insects whose body temperature changes with the ambient temperatures. These insects can easily survive between 12–21 degrees Celsius [1]. Temperature impact on their viability, fertility, developmental period, foraging activity, feeding, and breeding could easily be seen in laboratory stocks and distribution and populations dynamics under field conditions [2, 3]. Even Apart from temperature, humidity and rainfall, sunlight also plays a vital role in distribution of drosophilid species.

The family Drosophilidae encompasses 4,450 species distributed in 75 genera with two subfamilies, drosophilidae and steganinae [4]. The sub family drosophilinae is more diverse, distributed across 47 genera of which genus *Drosophila* is the largest with 1,213 species. The subfamily steganinae is a smaller one and is distributed across 28 genera of which genus *Leucophenga* is the largest with 256 species [4].

In India, more than 347 species are recorded which are spread across 27 genera, of which 58 species belongs to 8 genera of subfamily steganinae and rest 289 species of 19 genera are placed in subfamily drosophilinae (unpublished data).

2. Methods of collection

There are several methods in practice to collect fruit-flies from their natural habitats. Some of the methods are shown below:

2.1 Installing trap-bait

For setting-up traps one could use plastic bottles ranging in the volumes of 250 mL to 1 L. Fruit slices along with a pinch of yeast could be used as a bait. With the use of blade somewhere in the middle of the bottle a section could be carved out for food access. These traps could be hanged on orchard tree at a height 3–4 feet above the ground (**Figure 1**). The collection area must be damp and moist with minimal human interference. After 1–2 days traps must be recovered to collect flies. Same method could be applied to collect all kinds of *Drosophila* species. Some flies prefer to breed in the trash and stay close to ground, and some around the trees.

a)  b) 

Figure 1.
Trap bait: a) Transparent plastic bottles can be used as traps, a C-shaped window is made in the center of the bottle which serves as an entry point for the flies. Banana pieces placed inside the bottles attracts the flies. b) The bottle can be placed around small bushes/trees and can be collected next morning.

The methods could be improved based on little understanding of *Drosophila* species ecology (authors observations; unpublished).

2.2 Net swipe

There are many genera that are not attracted to regular traps and need to be captured from their natural food source (wild rotting fruits). Capturing such drosophilid species nets are more effective. The flies hovering over the rotten fruits or piles of organic wastes could be captured using this method (**Figure 2**).

2.3 Use of aspirator

This method is used when fly's numbers are low and also when investigator is familiar with the species identification and targeting particular species individuals. This method is more appropriate when flies are feeding, mating or resting on petals, leaves, fruits etc. Flies feeding on mushrooms and flowers are mostly collected using this method (**Figure 3**). Aspirator is also used to transfer flies from bait bottles or insect net to culture vials.

Some of the common species-specific baits are:

Banana: It is the most commonly used bait. It could be used to collect cosmopolitan *Drosophila melanogaster,* and other commonly found drosophilid species.

Figure 2.
Net sweeping: The modified insect nets are used for capturing drosophilids which are not attracted toward baits.

Figure 3.
Use of aspirator: The flies performing mating or resting over leaves, mushroom or fruits etc. are captured with help of aspirator.

Tomato: Adding yeast to it forms an attractive bait for *Drosophila suzukii* and *Drosophila busckii*.

Orange: It is also an attractive bait for *Drosophila suzukii* and *Drosophila busckii*.

Grapes: With added yeast grapes act as an attractive bait for species like *Drosophila busckii* and *Drosophila immigrians*.

Usage of vinegar in the baits for *Drosophila suzukii* helps in trapping more flies. Keeping baits in moisture free traps improves the collection yield.

3. Culture handling and maintenance

The culture handling and maintenance is required to be learned for making the collected sample to multiply several times and for making isolines and confirming the species's level identification.

3.1 Rearing and culturing *Drosophila*

For rearing and culturing the drosophilids, the collected samples need to be transferred from bait bottles to the narrow culture vials (with culture media) While shifting flies from the bottles to collection tubes the openings of the bottles must be kept under bright light directing the flies toward the light and thus making their transfer easy. For culturing, the flies are kept at room temperature (25°C). If room temperature is not adequate the culture tubes could be kept in the BOD (Biological Oxygen Demand) incubators. This would help the culture media to last longer until the drosophilid colony begins to establish.

3.2 Preparing *Drosophila* media

1. The commonly used media for various drosophilid stock maintenance includes Agar, yeast, Maize flour, Brown sugar, Nepagin, and Propionic acid.

2. To prepare fly media agar is added to the hot water. Following this yeast, maize flour, and sugar is added. After 20–30 minutes of cooking (with 1–2 boil) heat should be turned off. Once media temperature reaches to close to 50 degrees nepagin and propionic acid could be added. Throughout the preparation food needs to be constantly stirred.

3. Maize powder, brown sugar, and dried yeast are used as food. Yeast holds special nutritional value for drosophilids. While Nepagin is anti-fungal in nature and Propionic acid is a bactericidal and functions as preservative and increases the shelf life of food.

The food could then immediately be transferred to the sterilized vials or bottles. As soon as the media starts hardening the vials or bottles needs to be covered properly with the help of a cheese cloth. The food could be used after a day. The media tubes or bottles could also be stored in a cool place for 1–2 weeks for future use. *Other food recipes:*

There are almost 7–8 different food recipes for drosophilid species. The following link could be further explored for other recepies [5].

However, the standard procedure can be modified for the normal lab conditions and general use.

1. Cornmeal, sucrose, dextrose, yeast and 2-acid medium

This food recipe was described by Lewis in 1960. This uses the phosphoric acid which allows less propanoic acid usage without affecting the fungicidal effects. This recipe has also been used at Caltech since 1955.

2. Cornmeal, molasses and yeast medium

   This mixture has to be used fresh. This medium was used at Bloomington for several years. It went through certain modifications in order to prevent bacterial contamination.

3. Cornmeal, dextrose and yeast medium

   This recipe was first used by Brent and Oster in 1974. In order to reduce the chances of Leucon Stoc infection of cultures, sucrose was substituted by dextrose [5].

4. Identification (taxonomy)

In a biodiversity rich place, one could come across many small insect species. Identification becomes crucial in sorting a particular species from a pool of collection. Over the years, taxonomists came across various methods to identify species. Therefore, some common morphological diagnostic characters were listed out and based on these keys one could identify drosophilid species. However, closely related or sibling species are hard to distinguish with general morphological characteristics.

Drosophilids are usually small flies, ranging from 1.5–7.0 mm in length; yellowish, golden, brownish or blackish in color. They possess a number of the characteristics, such as red eyes and plumose arista. Body is shiny, often with stripes or spots on the thorax. Wings are hyaline or with black patches or marginal areas with dark lines. Abdominal tergite strip patterns (i.e. pigmentation) vary from species to species (dark or light bands or spots in 2–6 tergites). In some species, sexual dimorphism is clear by wing patch or presence of sex-comb on legs.

However, species are identified by their male genital organs (periphallic & phallic) possessing structural compatibility features [6]. These genital organelle structures are species specific. Example, the two subfamilies viz., drosophilinae and steganinae are differentiated on the basis of the distance between proclinate orbital setae and inner vertical setae from posterior reclinate setae. In subfamily drosophilinae the distance of proclinate orbital seate from posterior reclinate setae is less compared to its the distance of inner vertical setae. While in subfamily Steganinae the distance of proclinate orbital setae from posterior reclinate setae is more than compared to its distance from inner vertical setae.

Flies are too small in size to be observed with naked eyes. Hence magnification is required and this could be achieved with a good hand lens, or a wide-field binocular microscope, or a stereo zoom microscope, or a compound microscope. On a white background pictures emerge better (**Figure 4**).

For identification, anesthesia could be given to flies. This makes flies unconscious. Ether or Carbon dioxide gas could be used as anesthesia. Flies sensitive to ether or CO2 cold treatment is an option.

Generally, females possess larger body size and have swollen abdomen than males. In some cases of males' 5th and 6th tergites are pigmented whereas others have wing patches. In a few genera of drosophilids males also possess sex comb on their fore legs. The posterior end of males body is pointed whereas in females it is pointed (due to the presence of their ovipositor and female genital organ). The male genital organs are species-specific and differ from species to species. Dissection of genital organs is used for conformation of species. Dissection of male genitals is usually done with the help

a) 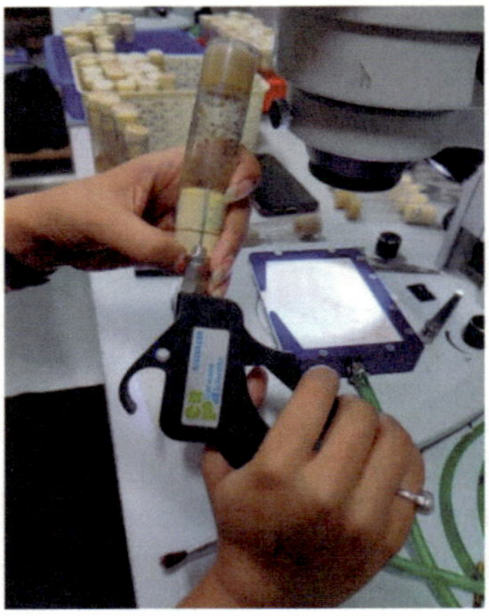 b) 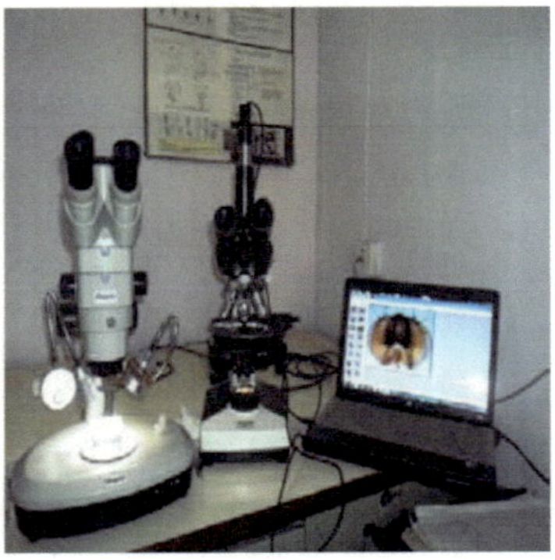

Figure 4.
Identification: a) after collection of the flies, they are given anesthesia (CO_2) which gives 30 minute time to sort the male and female. b) Identification can also be done by dissecting the male genitalia.

of a needle. It requires washing the tissues in 10% KOH approximately at 100 °C for several minutes. This opens the intact reproductive plates and helps investigators to collect additional details. Few drops of glycerol can be added for better resolution.

4. Morphological study of drosophilid

Like in other arthropods, the adult drosophilids body is comprised of three major parts i.e., head, thorax and abdomen. Drosophilids have one pair of wings and halters each. Wings help in flight whereas halters help as balancing organs.

4.1 Head

The head of drosophilids has distinguishing parts like ocellar triangle, post-ocellar setae, inner and outer vertical setae, fronto-orbital plates, cibarium and arista. Arista is a distinctive character as it possesses varying dorsal and ventral branches in different genus. The genus *Drosophila* possesses three elongated orbitals with proclinate setae inserted into the anterior most part. The ratio, length and placement of orbitals, ocellar, vertical and post-ocellar setae is used for the identification of different genera of drosophilids (**Figure 5**). In genus *Chymomyza* the anterior reclinate setae is present in front of the proclinate setae whereas in genus *Liodrosophila* anterior reclinate setae present is small in size. Facial carina is also an important diagnostic characteristic in the species. The chaetotaxy and color of the palps is used to identify the sibling species [7]. The setae on either side of the face known as vibrissae and sub-vibrissae are important for the identification of species. The number of anterior & posterior sensilla and sensillacampaniformia of the cibarium are important diagnostic characteristics for the identification of the species.

4.2 Thorax

The thorax has three main segments: prothorax, mesothorax and metathorax. Mesothorax is significantly enlarged which aids the wings, while prothorax and metathorax are generally reduced. Most of the dorsal surface of the mesothorax is covered with mesonotum. There are numerous regular and irregular rows of

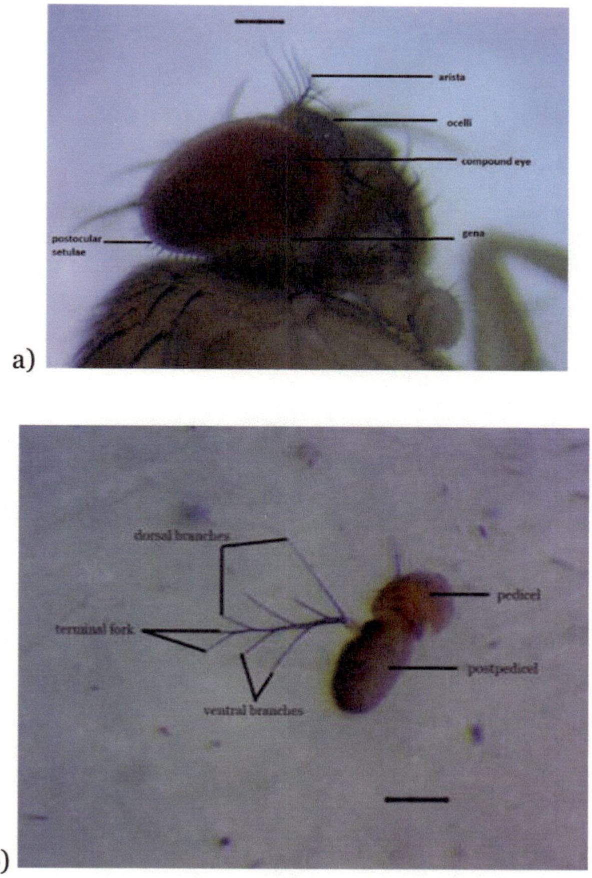

Figure 5.
Head: a) D. suzukii, showing major portions of head region. b) Arista and its parts.

acrostichalsetulae and numeral pairs of dorsocentral setae present on the mesonotum. The number and position of the Acrostichalsetulae is taxonomically important for differentiation of the species (**Figure 6**). Majority of the *Drosophila* species have six or eight regular acrostichal rows. While most of the *Scaptomyza* species possess two or four rows. However, other genres are characterized by ten or more irregular rows. In a few species of mycophagous *tripunctata and testacea* group a set of enlarged acrostichals are located more anteriorly near to the transverse suture known as presutural setae and are also used for identification [7]. The length and orientation of basal and apical scutellar setae (present on the scutellum) are also taxonomically important for the species identification. Number and length of the katepisternal setae present on katepisternum are also significant for differentiation of the species and species group.

4.3 Legs

Legs in drosophilids are divided into coxa, trochanter, femur, tibia and tarsus. Tarsus have 5 tarsal segments. Color and arrangements of bristles (chaetotaxy) of male foreleg and relative length of first tarsus is important for distinguishing traits among different species. Their presence and the number of spines on hind leg, tibia and tarsomere (genus *Impatiophila)* both are considered for the identification of closely related species [8]. In some species such as *Liodrosophila angulata* a number of taxonomically important spines are present on the femur of the foreleg.

a) 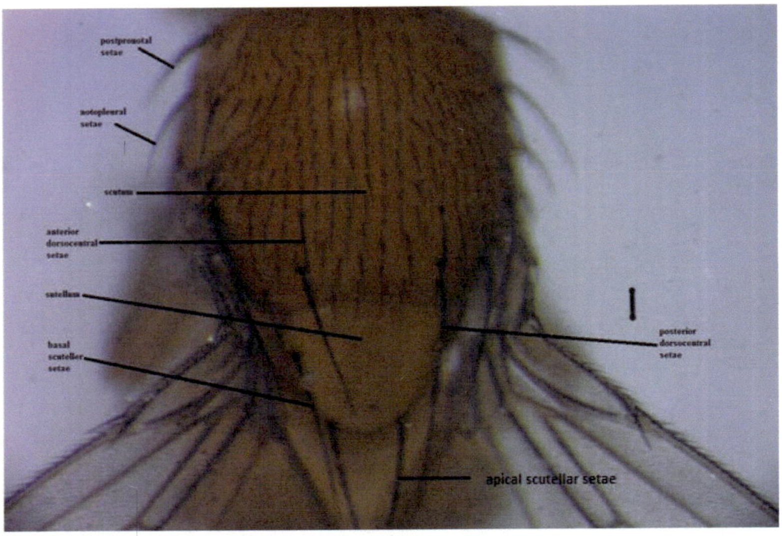

b) 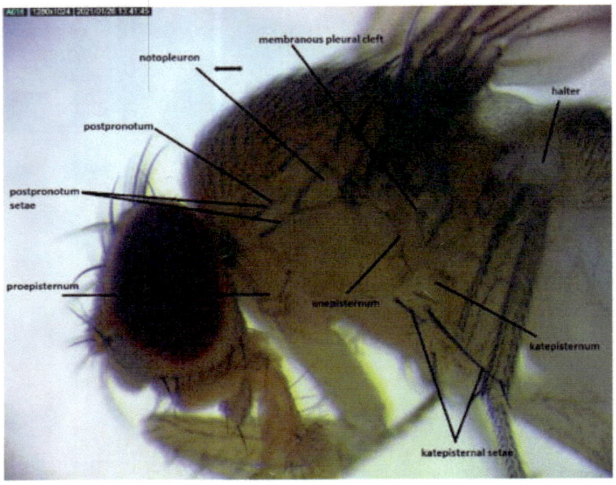

Figure 6.
Thorax: D. suzukii a) dorsal b) lateral.

4.4 Sex Combs

Some species of *melanogaster* and *obscura* group are characterized on the basis of sex combs on the male foreleg. The numbers of sex combs are taxonomically important for the distinction of the different species. *Drosophila* males use their sex combs to grasp the females' abdomen and genitalia. They also use them to spread their wings prior to copulation.

4.5 Wings

Wings are attached to the mesothorax of the abdomen. In many drosophilids wings possess spotted patterns. Also,hylanine or fuscous are taxonomically important. Although there is a consistent pattern on wing's venation but in some taxa additional cross veins could be present eg *planitibia* species group. Different types of cross-veins (bm-cu, dm-cu, r-m and cuA2), subcostal break, humeral break, wing cells (bm, dm, cup), and wing veins (CuA1, A1, R1, R2 + 3, R4 + 5, M1) occur in the wings (**Figure** 7). In genus *Impatiophila* the setae of the middle row on the

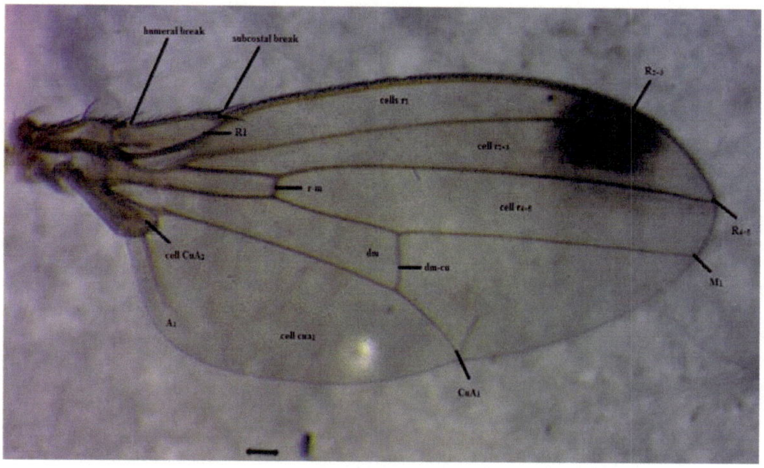

Figure 7.
Wing: Abbreviations; h = humeral, hum brk = humeral break, R1 = anterior branch of radium, C1 s = apical seta(e) on 1st costal section, sc brk = subcostal break, r2 + 3 = second + third radial, r4 + 5 = fourth + fifth radial, M1 = 1st posterior (sectorial) branch of media, dm-cu = discal medial-cubital, CuA2 = 2nd anterior branch of cubitus, CuA1 = 1st anterior branch of cubitus, A1 = 1st branch of anal vein, r-m = radial-medial. (source; terminology by professor M.J Toda).

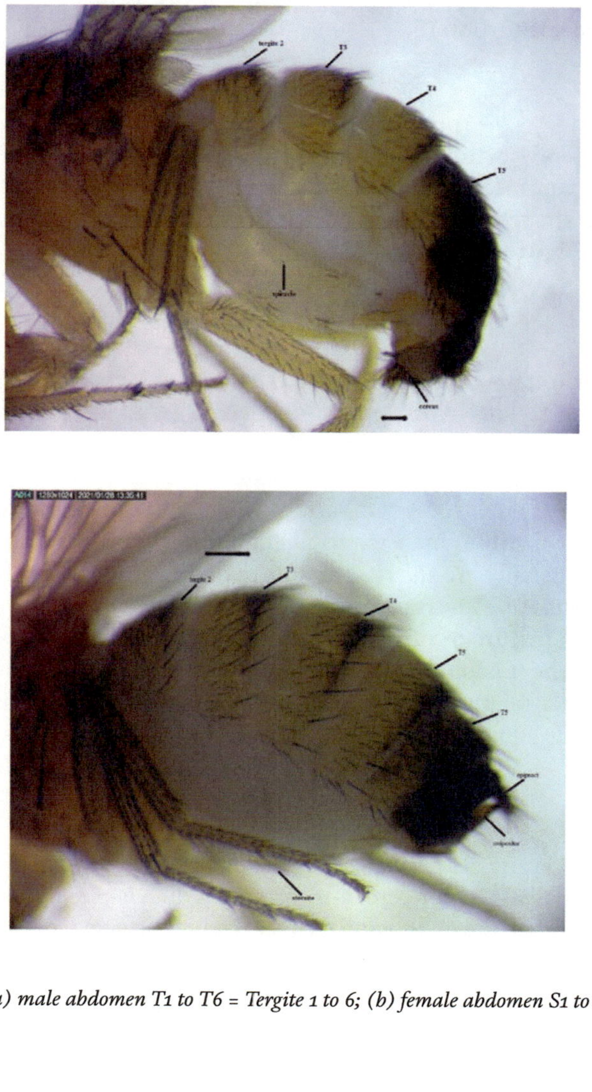
a)
b)

Figure 8.
Abdomen: D. suzukii (a) male abdomen T1 to T6 = Tergite 1 to 6; (b) female abdomen S1 to S6 = Sternite 1 to 6.

Figure 9.
Species level distinguishing characters.

second costal section is considered important for the differentiation of the species into different species group [8].

4.6 Abdomen

The dorsolateral portion of abdomen is known as tergite and it is segmented and chitinized. The ventral portion is called sternite. It is generally hairy, chitinized and quadrilateral in shape. The length and width of the sternite is important for species identification (**Figure 8**).

4.7 Terminalia

The male terminalia possess many internal and external characters useful for species characterization. In drosophilids the male genitalia exhibit speedy and divergent evolution while female genitalia are thought to evolve slowly among closely-related species. In drosophilids female copulatory structures have been claimed to be mostly invariant compared to male structures. In most animal species with internal fertilization, male external genitalia are the most rapidly evolving organs and they usually are the first organs to diverge morphologically following by speciation. Because of their rapid evolution and species-specificity, their illustration is a common feature of taxonomic literature to discriminate between closely-related species. The morphology of male genitalia can differ dramatically even within very closely related animal species. The male terminalia is further divided into epandrium and hypandrium.

4.8 Epandrium

In males, 9th tergite is known as epandrium and it possesses a number of characteristics such as pair of cerci and surstyli which is present on the posterior and posteroventral of the epandrium respectively. The size, color, shape, morphology and number of setae on the cerci are significant for the distinction of species. Surstyli bears a number of distinct prensistae. The numbers of prensistae vary from species to species and are important for the species identification.

4.9 Hypandrium

The genitalia are the structure linked with 9th sternite or Hypandrium and possesses aedeagus, basal processes of the aedeagus, parameres, and gonopods on it. The posterolateral portion of the hypandrium is known as gonopods. Posterolateral to the gonopods are paraphyses which possess a number of setae. The aedeagus is placed centrally with respect to the rest of hypandrium.

The major distinguishing character at species level are the presence and numbers of sex combs and bristles. Different types of spots present on wings and abdomen, and lining present on thorax of different drosophilids etc. are the characters that provide cues to identify flies (**Figure 9**).

5. What makes *Drosophila* a great model organism?

5.1 *Drosophila* as a model organism

The different characteristics of *D. melanogaster* make it an ideal model organism, which are following:

Smaller Size and Short lifespan

Shorter life span facilitates large quantities of flies to be produced in a short time.

Minimal culturing requirements

Due to the smaller size and minimal requirements, *Drosophila* can be cultured and tested in limited resources.

Genetic manipulation

The fly genome has been sequenced and well characterized. It has 100+ years of literature available. Besides this it has four pairs of chromosomes only which makes an ideal system to do genetic crossing and gene editing simpler.

5.2 Basic research

As drosophilids are versatile, low maintenance and non-harming model organisms, they are used in all fields of life sciences like Genetics, Biotechnology, Cancer biology, Genomics, Reproductive biology, Developmental biology, Micro chemical studies, ecology and much more. For more than a century, the low cost, rapid generation time, and excellent genetic tools have made the fly indispensable for basic research. Also, the recent advancements in the field of molecular tools have allowed the organism to be used more efficiently. From human disease modeling to the dissection of cellular morphogenesis and to behavior and aging, the current usage of flies greatly influences fly research. However, this field remains vibrant and exciting, with labs using flies in drug discovery, bioengineering, regenerative biology, and medicine. The future use of fruit flies as a model organism in research is bright.

6. Stock centres

Bloomington stock centre maintains various fly stocks including aberrations, balancers, deficiencies, duplications, clonal, chemically induced mutations, human disease model, mapping, teaching stocks, wild-type lines, transgenes and various other stocks. Further details about the stock centre could be found at site dedicated to stock centre [9].

Here is the list of other stock centres around the globe:

1. Kyoto Stock Centre, Kyoto Institute of Technology, Kyoto, Japan.

2. Harvard Medical School, Boston, MA, USA.

3. FlyORF, University of Zurich, Zurich, Switzerland.

4. NIG-FLY, National Institute of Genetics, Mishima, Japan.

5. THFC, Tsinghua University, Beijing, China.

6. Vienna Drosophila Resource Centre (VDRC), Vienna, Austria

7. Fly maintenance stuff

A detailed information about fly maintenance materials and accessories are valuable at several commercial vendors. The provided links provide access to it [10].

Important websites for Identification of drosophilids

Taxonomic information database for the world Drosophilidae:

DrosWLD Species (Taxonomic information database for world species of Drosophilidae, maintained by Masanori J. Toda) https://bioinfo.museum.hokudai.ac.jp/

Japan *Drosophila* Database: JDD http://www.drosophila.jp/jdd/index_en.html

7. Conclusions

This chapter highlights the basic clues to identify different species in the field and lab along with learning distinguishing characteristics of males and females easily and effectively.

8. Recommendations

These protocols will act as baseline data for handing and maintaining drosophilids for young taxonomist. As these processes are easy ones so these could be used at graduation level to let students get familiar with its taxonomy.

Acknowledgements

The authors wish to offer cordial thanks to Dr. Subhash Rajpurohit, Associate Professor, Division of Biological and Life Sciences, School of Arts and Science, Ahmedabad University, Gujrat, India for his careful reading of the manuscript and his valuable suggestions for this study. This work is not financially support from any agency except the publisher.

Conflict of interest

The authors declare no conflict of interest.

Acronyms and abbreviations

Wing h	humeral
hum brk	humeral break
R1	anterior branch of radium
C1 s	apical seta(e) on 1st costal section
sc brk	subcostal break
r2 + 3	second + third radial
r4 + 5	fourth + fifth radial
M1	1st posterior (sectorial) branch of media
dm-cu	discal medial-cubital
CuA2	2nd anterior branch of cubitus
CuA1	1st anterior branch of cubitus
A1	1st branch of anal vein
r-m	radial-medial. (Source; Terminology by Professor M.J Toda). Abdomen: T1 to T6 = Tergite 1 to 6; S1 to S6 = Sternite 1 to 6;

Author details

Pragya Topal[1], Divita Garg[2] and Rajendra S. Fartyal[1*]

1 Department of Zoology, HNB Garhwal University, Srinagar, Uttrakhand, India

2 Division of Biological and Life Sciences, School of Arts and Sciences, Ahmedabad University, Ahmedabad, Gujarat, India

*Address all correspondence to: fartyalrs@gmail.com

References

[1] Hoffmann, A. A. Physiological climatic limits in Drosophila: Patterns and implications. Journal of Experimental Biology; (2010). 213(6), 870-880. https://doi.org/10.1242/jeb.037630

[2] Clemson, A. S., Sgrò, C. M., & Telonis-Scott, M. Thermal plasticity in *Drosophila melanogaster* populations from eastern Australia: Quantitative traits to transcripts. Journal of Evolutionary Biology; (2016). 29(12), 2447-2463. https://doi.org/10.1111/jeb.12969

[3] Moghadam, N. N., Ketola, T., Pertoldi, C., Bahrndorff, S., & Kristensen, T. N. Heat hardening capacity in *Drosophila melanogaster* is life stage-specific and juveniles show the highest plasticity. Biology Letters; (2019). 15(2), 20180628. https://doi.org/10.1098/rsbl.2018.0628

[4] "Welcome to T a x o D r o s." https://www.taxodros.uzh.ch/ (accessed Mar. 01, 2021).

[5] "Bloomington Drosophila Stock Center: Indiana University Bloomington." https://bdsc.indiana.edu/information/recipes/caltechfood.html (accessed Mar. 01, 2021).

[6] Bächli, G. Family Drosophilae. In: Papp, L. and Darvas, B. (eds). Contributions to a manual of Palearctic Diptera. III. Higher Brachteera. Science Herald; 1998.Vol –3 pp 503-513.

[7] Markow, T.A., & O'Grady, P. Drosophila: a guide to species identification and use. Elsevier (2005).

[8] Fu, Z., Toda, M. J., Li, N.-N., Zhang, Y.-P., & Gao, J.-J. A new genus of anthophilous drosophilids, Impatiophila (Diptera, Drosophilidae): Morphology, DNA barcoding and molecular phylogeny, with descriptions of thirty-nine new species. Zootaxa; (2016). 4120(1), 1-100. https://doi.org/10.11646/zootaxa.4120.1.1

[9] "Bloomington Drosophila Stock Center," *Bloomington Drosophila Stock Center*. https://bdsc.indiana.edu/index.html (accessed Mar. 01, 2021)

[10] "Flystuff." https://flystuff.com/ (accessed Mar. 01, 2021).

Section 2

Distinctive Diptera

Chapter 4

Diversity of Tephritidae and Agromyzidae (Diptera: Brachycera) in Flower Heads of Asteraceae in the Chaco

Manoel A. Uchoa, Anderson S. Fernandes and Jimi N. Nakajima

Abstract

The Chaco is an international biome, connecting four countries: Paraguay (230,000 km^2), Bolivia (90,000 km^2), Argentina (520,000 km^2), and Brazil (Mato Grosso do Sul state (MS), with around 9,000 km^2 and in the middle of South America. Brazilian Chaco is restricted to Porto Murtinho region, MS. The daisies (Asteraceae) with near 24,000 species worldwide is characterized by herbs and shrubs that coevolved with several taxa of endophagous insects: dipterans Agromyzidae, Ceciidomyidae and Tephritidae; Coleoptera (Apionidae), Hemiptera (Miridae), Lepidoptera (Blastobasidae, Gelechiidae, Pterophoridae, Pyralidae, and Tortricidae) and the parasitoids of this endophagous insects, which found in the daisies's flower heads ideal conditions for food, breeding site and shelter. The Neotropical florivorous flies are the Agromyzinae (Agromyzidae), and Tephritinae (Tephritidae), which in their larval stage feed on Asteraceae inflorescences. To report the species of florivore flies, their host plants and parasitoids in flower heads of Asteraceae from the Brazilian Chaco, we sampled inflorescences of 25 species (± 500 flower heads/species) that were kept in containers to the emergence of the florivorous flies or their parasitoids sampled in the three phytophysiognomies. The adult insects after 48 hours of their emergence were fixed in 80% ethanol for later identification. A total 25 species of Asteraceae were evaluated in the Brazilian Chaco, being collected 17,000 flower heads. Nine tribes of two Asteraceae subfamilies were sampled, from which 15 species of florivorous flies were recovered. We found 5 genera with 9 of Tephritinae (Tephritidae), 6 species of *Melanagromyza* (Agromyzinae, Agromyzidae), and 104 parasitoids (Hymenoptera) of the florivorous flies.

Keywords: Florivory, *Melanagromyza* spp., Steppe Savannas, Tephritinae, weed biocontrol

1. Introduction

The florivorous flies are the most importante guild of insects feeding on daisis's flowers in Brazil. They belong to the subfamilies Agromyzinae (Agromyzidae), and Tephritinae (Tephritidae). Those flies, in their larval stage, feeds in Asteraceae inflorescences (flower heads).

Asteraceae is the bigest family of Angiospermae with around 24,000 species of plants worldwide, being characterized by herbs and shrubs that coevolved with several taxa of endophagous insects, which found in their flower heads ideal conditions for food, breeding site and shelter. As pointed out by Lewinsohn [1], several other taxa on Insecta has been reported on the Asteraceae flower heads, beyound florivore flies, such as Apionidae (Coleoptera), Miridae (Hemiptera), Blastobasidae, Gelechiidae, Pterophoridae, Pyralidae, and Tortricidae (Lepidoptera), as well their parasitoids, mainly Hymenoptera.

In Neotropical Region, the species of Asteraceae, depending on the biome evaluated, use to be the first or sencond plant Family in the rank of species richness, as pointed out for the Atlantic Forest on Espirito Santo state, Brazil [2]. The studies about the florivore insects are important for both: to expand scientific knowledge, as well as to solve environmental problems in food production systems. By now, at least 38 species and their subspecies of invasive Asteraceae have been the target of biological control programs around the world. About 21 species of Tephritidae have been manipulated to control Asteraceae species. The most significant biological control programs for such plants with florivorous flies are underway in Australia, Canada, USA, New Zealand and South Africa [3], and West Africa, where the stem-gall tephritid *Cecidocares connexa* (Macquart 1848) have been upplied on the biological supperession, against the invasive Asteraceae *Chromolaena odorata* (L.) R. M. King & H. Rob. [4]

Chaco is an international biome, distributed by four South American countries: Argentina, Paraguay, Bolivia and Brazil. The total area of this biome is spread over the four above countries: Paraguay (230,000 km^2), Bolivia (90,000 km^2), Argentina (520,000 km^2), and Brazil (Mato Grosso do Sul = MS, with around 9,000 km^2), in the center of South America. The word Chaco, derives from the indigenous term: "Quechua chaku" – that mean hunting territory or steppe savanna. The Brazilan Chaco is restricted to the municipality of Porto Murtinho, located in the south of the Pantanal, state of MS, where it is characterized as the humid Chaco. This sedimentary plain is covered by natural vegetation of Steppe Savanna and occupy the central region of South America, called Gran Chaco or simply Chaco. It's covered by the most extensive continuous dry forest in the American continent, and represents the only one subtropical dry forest of the world. The Chaco region, with approximately 1,000,000 km^2 (Gran Chaco), covers North Argentina (46%), West of Paraguay (32%), Southeast of Bolivia (15%), and a significant portion (7%) in the Midwest of Brazil (Porto Murtinho, Mato Grosso do Sul), separated from the biome matrix by the Paraguay river [5, 6].

The Chaco Biome is home to a great diversity of environments: mountains, savannas, floodplains, swamps, wetlands and saltpans, with a great extension and diversity of flora and fauna. The ecosystems that are part of the Chaco have high rates of endemism and diversity of plant and animal species, compared to other arid, semi-arid and sub-humid environments. Nowadays, Chaco faces great anthropic pressure, due to the advances of agriculture that has increased, exploiting its phyto ecological regions. The high biodiversity of the Brazilian Chaco, and its rates of endemismo [5, 6], suggest that native fruit plants are potential hosts for some frugivorous and florivorous fly species there.

Tephritidae is the most important, and the second family in species richness (behind Cecidomyiidae) of the phytophagous Diptera, with around 5,000 species, being known mainly due the economic importance of their pest species upon fruit and vegetables [7]. But there are another face of these flies: those species that feed on plants, but don't cause economic loss. By other hand, they are of great economic importance due to their phytophagy, destroying seeds of undesirable Asteraceae species in agrosilvipastoral systems. These florivorous flies (Tephritinae), the bigest

Subfamily of Tephritidae - near 2,000 species, were subject of our research [8], and will be presented in this e-book chapter. We evaluated the diversity of its host Asteraceae, species richness and abundance of florivore's fly species in each species of Asteraceae evaluated, as well as the co-occurrence of other taxons of florivorous dipterans and their respective parasitoids in the Brazilian Chaco.

2. The mainly aims of this research

1. To Report the species of florivore flies (FF), their host plants, and their parasitoids, in flower heads of Asteraceae from the Brazilian Chaco. 2. To quantify abundance, species richness and diversity of the FF in three phytophysiognomies in the Brazilian Chaco.

3. Materials and methods

Sampling was carried out in the municipality of Porto Murtinho-MS, Brazil, in three different phytophysiognomies of the Brazilian Chaco. Flower heads of Asteraceae were collected at: Eldorado Farm (21 ° 42'20.9 "S 57 ° 47'45.6" W, altitude 82 m); Santa Carmem Farm (21 ° 50'23.2 "S 57 ° 49'13.6" W, altitude 78 m), and in a transect along the Highway MS-457 (altitude 124 m).

At the Eldorado farm, the predominant phytophysiognomy is of the type Wooded Steppe Savanna (WSS), and at the farm Santa Carmem the type is Park Steppe Savanna (PSS), subtype locally named *Carandazal*), with formation of Riparin Forest. Carandazal is characterized by the vegetation with predominance of the *Copernicia alba* Morong palm. In these two locations the transects were of approximately 1 km in open field, with a sampling effort of one person during 2 h. On Highway MS-457 trail, the predominant phytophysiognogmy is of the Grassy-Wood Savanna (GWS). The collections were carried out along the edges of this trial, starting from BR-267 (21 ° 44'51.4 "S 57 ° 33'44.1" W) towards the *Parque Municipal Cachoeira do APA* (= APA Waterfall Municipal Park). This last trail covered about 2 km.

The flower heads of all Asteraceae species found in the field were collected, preferably 200 in pre-anthesis per species. The flower heads were inserted in plastic bags, identifying the species of plant and the area. The collection carried out monthly and during from May 2017 to April 2018. Exsiccates from all Asteraceae evaluated were prepared.

The climate of the region in the Brazilian Chaco is considered tropical Aw by the Köppen classification, with hot and rainy summer and dry and milder winter. The rainfall varies between 1,100 mm and 1,800 mm. The vegetation in the Chaco is made up of shrubs, deciduous, microphiles and spinach. The soil is highly saline, with little water drainage, which is why in rainy season it generates temporary floods characteristic of the Pantanal.

The Asteraceae inflorescences collected were taken to the Laboratory of Systematic and Taxonomy of Tephritidae (LabTaxon), Universidade Federal da Grande Dourados (UFGD), where the flower heads were counted and placed in 500 ml plastic cups, with the juxtaposed openings forming a cage. The cups were attached with adhesive tape, forming a closed container that made it possible to contain and obtain species of florivorous flies and their parasitoids. After emergence of florivorous flies and/or their parasitoids, they remained alive for a period of 48 hours to acquire the chromatic pattern of the body and wings. They were subsequently conserved in 92% ethyl ethanol. After 20 days without any fly's emergence, inspections on the containers were stopped, and the remaining material discarded.

The exsiccates of the sampled Asteraceae were sent for identification of the species by Professor Dr. Jimi Naoki Nakajima (Universidade Federal de Uberlândia), specialist in Asteraceae. The duplicates of the identified Asteraceae were deposited at the CGMS Herbarium of the Biodiversity Museum-FCBA/UFGD in Dourados-MS, as the specimens voucher.

Statistical analyzes: Infestation rates were calculated using the equation: Number of FF/Number Flower heads x 100, being N the number FF (= number of Florivorous flies) divided by the Number of Flower heads (total of Flower heads/Asteraceae species). Were analysed the diversity in each phytohysiognomy. The diversity index (Shannon-Weaner) (H), Species Richness Index (Margalef), (Alpha), and Uniformity or Equitability Index (E) were calculated.

4. Results

Twenty-five species of Asteraceae of two subfamilies (Cichorioideae and Asteroideae), and 9 tribes (Vernonieae, Senecioneae, Astereae, Inuleae, Plucheae, Neurolaeneae, Heliantheae, Targeteae, Eupatorieae) were sampled in three phytophysiognomies: Grassy-Woody Steppe Savana, Park Steppe Savanna, and Wooded Steppe Savanna from the Brazilian Chaco at Porto Murtinho-MS, Brazil (**Table 1**).

A total 472 adults of florivorous flies from two families and 15 species (9 of Tephritidae and 6 of Agromyzidae) were recovered, being 163 agromyzids (*Melanagromyza*), and 309 tephritids from nine species were recovered from 15 species Asteraceae species. The Tephritid species belong to five genera: *Cecidochares* Bezzi 1910; *Dictyotrypeta* Hendel 1914; *Tomoplagia* Coquillett 1910; *Trupanea* Schrank 1795, and *Xanthaciura* Hendel 1914. We found nine morphospecies of Tephritinae. All specimens of Agromyzidae are grouped in a single genus: *Melanagromyza* Hendel 1920. The recovered parasitoids are (n = 104) Hymenoptera await identification by specialists. Herein, trophic interactions are reported between florivorous flies with 15 species of Asteraceae in the Brazilian Chaco (**Table 2**).

Among the 16 species of Asteraceae that host florivorous flies or their parasitoids, five were associated only with agromizids: *Bidens gardneri*, *Bidens pilosa*, *Acmella grisea*, *Wedelia brachycarpa* and *Cyrtocymura scorpioides*. Six hosted only tephritids: *Chromolaena ivifolia*, *Praxelis clematidea*, *Pectis odorata*, *Porophyllum angustissimum*, *Lepidaploa remotiflora* and *Lepidaploa remotiflora*. Four Asteraceae species hosted species of both florivorous flies (Tephritine and Agromyzinae): *Aspilia elata*, *Chromolaena margaritensis*, *Dimerostemma grazielae* and *Porophyllum ruderale*. Parasitoids emerged from the florivorous flies feeding on 11 Asteraceae species (**Table 2**).

Herein, for the first time we present the several associations of florivorous fly species with their Asteraceae (flower heads) host species in the Brazilian Chaco. The trophic interactions established here were the following: *Chromolaena margaritensis* was colonized by *Xanthaciura* sp.1; *Aspilia elata* by *Dictyotrypeta* sp.1, and *Dictyotrypeta* sp.2; *Dimerostemma grazielae* by *Trupanea* sp.1; *Pectis odorata* by *Trupanea* sp.1; *Lepidaploa remotiflora* by *Tomoplagia minattai*. From *Lessingianthus niederleinii* flower heads emerged adults of *Tomoplagia matzenbacheri*. From the flower heads of *Chromolaena ivifolia* occurred the highest abundance of florivorous flies (166). This Asteraceae species was infested only by species of tephritids. From *Aspilia elata* flower heads, were obtained greatest abundance of agromyzids (62) (**Table 2**).

The highest rates of infestation by florivorous flies in Asteraceae in the Brazilian Chaco were reported in the following species: *Acmella grisea* (34%); *Aspilia elata*

Asteraceae taxa (subfamilies, tribes and species)	Host status	Environments (Phytophysiognomies)
Cichorioideae		
Vernonieae *Cyrtocymura scorpioides* (Lam.) H. Robins.	Host	Grassy-Woody Steppe Savanna (GWSS)
Lepidaploa remotiflora (Rich.) H. Robins.	Host	Wooded Steppe Savanna (WSS) Park Steppe Savanna (PSS) Grassy-Woody Steppe Savanna (GWSS)
Lessingianthus niederleinii (Hieron.) H. Robins.	Host	Grassy-Woody Steppe Savanna (GWSS)
Lessingianthus rubricaulis (Bonpl.) H. Robins.	Nonhost	Grassy-Woody Steppe Savanna (GWSS)
Asteroideae		
Senecioneae *Erechtites hieracifolia* (L.) Raf.	Nonhost	Wooded Steppe Savanna (WSS)
Astereae *Conyza bonariensis* (L.) Cronquist	Nonhost	Wooded Steppe Savanna (WSS)[a]
Inuleae *Pluchea quitoc* DC.	Nonhost	Wooded Steppe Savanna (WSS) Grassy-Woody Steppe Savanna (GWSS)
Plucheae *Pterocaulon virgatum* (Lam.) DC.	Nonhost	Grassy-Woody Steppe Savanna (GWSS)
Neurolaeneae *Calea rupicola* Chodat	Nonhost	Grassy-Woody Steppe Savanna (GWSS)
Heliantheae *Acmella grisea* (Chodat) R.K. Jansen	Host	Grassy-Woody Steppe Savanna (GWSS)
Aspilia elata Pilg.	Host	Park Steppe Savanna (PSS) Grassy-Woody Steppe Savanna (GWSS)
Aspilia montevidensis (Spreng.) Kuntze	Nonhost	Grassy-woody Steppe Savanna (GWSS)
Bidens pilosa L.	Host	Grassy-Woody Steppe Savanna (GWSS)
Bidens gardneri Baker	Nonhost	Grassy-Woody Steppe Savanna (GWSS)
Dimerostemma grazielae H. Rob.	Host	Wooded Steppe Savanna (WSS) Park Steppe Savanna (PSS) Grassy-Woody Steppe Savanna (GWSS)
Wedelia brachycarpa Baker	Host	Wooded Steppe Savanna (WSS) Park Steppe Savanna (PSS) Grassy-Woody Steppe Savanna (GWSS)
Targeteae *Pectis odorata* Griseb.	Host	Wooded Steppe Savanna (WSS) Park Steppe Savanna (PSS) Grassy-Woody Steppe Savanna (GWSS)
Porophyllum angustissimum Gardner	Host	Park Steppe Savanna (PSS) Grassy-Woody Steppe Savanna (GWSS)
Porophyllum ruderale (Jacq.) Cass.	Host	Wooded Steppe Savanna (WSS) Park Steppe Savanna (PSS) Grassy-Woody Steppe Savanna (GWSS)
Eupatorieae *Austroeupatorium inulifolium* (Kunth) R.M. King & H. Rob.	Nonhost	Grassy-Woody Steppe Savanna (GWSS)[b]
Campuloclinium macrocephalum (Less.) DC.	Nonhost	Grassy-Woody Steppe Savanna (GWSS)
Chromolaena ivifolia (L.) R. M. King & H. Rob.	Host	Grassy-Woody Steppe Savanna (GWSS)

Asteraceae taxa (subfamilies, tribes and species)	Host status	Environments (Phytophysiognomies)
Chromolaena margaritensis (Hassl.) R. M. King & H. Rob.	Host	Wooded Steppe Savanna (WSS) Park Steppe Savanna (PSS)c Grassy-Wood Steppe Savanna (GWSP)
Praxelis clematidea (Griseb.) R. M. King & H. Rob.	Host	Wooded Steppe Savanna (WSS) Grassy-Woody Steppe Savanna (GWSS)
Urolepis hecatantha (DC.) R. M. King & H. Rob.	Nonhost	Wooded Steppe Savanna (WSS) Grassy-Woody Steppe Savanna (GWSS)

[a] *Wooded Steppe Savanna (WSS) = Fazenda Eldorado (= Eldorado Farm);*
[b] *Grassy-Woody Steppe Savanna (GWSS), Access to Rio Apa waterfall (Border with Paraguay), and Transition to the Brazilian Cerrado;*
[c] *Park Steppe Savanna (PSS) = Fazenda Santa Carmen (= Santa Carmen Farm), "Carandazal Phytophysiognogmy".*

Table 1.
Status of Asteraceae species for florivorous flies Tephritinae (Tephritidae, and/or Melanagromyza *Hendel 1920, Agromyzinae: Agromyzidae) in three Phytophysiognomiesof the Brazilian Chaco.*

Asteraceae Taxa	N° of Flower Heads	Agromyzidae	Tephritidae	* I.L.FF/F.H. (%)	Parasitoids (Hymenoptera)
Tribo					
Astereae *Conyza bonariensis*	50	0	0	0	0
Coreopsideae					
Bidens gardneri	128	3	0	2.34	0
Bidens pilosa	85	6	0	7.05	0
Eupatorieae					
Austroeupatorium inulifolium	2,500	0	0	0	0
Campuloclinium macrocephalum	211	0	0	0	0
Chromolaena ivifolia	1,100	0	166	15.09	21
Chromolaena margaritensis	2,299	12	18	1.30	5
Praxelis clematidea	424	0	1	2.12	38
Urolepis hecatantha	1,245	0	0	0	0
Heliantheae					
Acmella grisea	139	48	0	34.53	0
Aspilia elata	320	62	29	28.43	4
Aspilia montevidensis	73	0	0	0	0
Dimerostemma grazielae	1,019	23	3	2.55	12
Wedelia brachycarpa	256	3	0	1.17	0

Asteraceae Taxa	N° of Flower Heads	Agromyzidae	Tephritidae	* I.L.FF/F.H. (%)	Parasitoids (Hymenoptera)
Inuleae					
Pluchea quitoc	290	0	0	0	5
Pterocaulon alopecuroides	1,280	0	0	0	0
Neurolaeneae					
Calea rupicola	210	0	0	0	0
Senecioneae					
Erechtites hieracifolius	50	0	0	0	0
Tageteae					
Pectis odorata	1,123	0	17	1.51	2
Porophyllum angustissimum	925	0	16	1.72	4
Porophyllum ruderale	1,343	3	51	4.02	7
Vernonieae					
Cyrtocymura scorpioides	120	3	0	2.50	0
Lepidaploa remotiflora	995	0	2	0.20	2
Lessingianthus niederleinii	585	0	6	1.02	4
Lessingianthus rubricaulis	250	0	0	0	0
Total	17,000	163	309		104

[1]*I.L.FF/F.H. = Infestation Level of Florivorous Flies by the number of Flower heads in Asteraceae species.*

Table 2.
Abundance Florivorous flies and parasitoids (Hymenoptera) associated with flower heads of Asteraceae species in three phytophysionomies in the Brazilian Chaco (at Porto Murtinho-MS) (May 5, 2017 to April 5, 2018).

(28%), both from the tribe Heliantheae, and from *Chromolaena ivifolia* (15%) (Eupatorieae). The species of *Trupanea* (Tephritinae) showed the highest number of trophic interactions (6) with Asteraceae. *Trupanea* was also the only Tephritidae genus associated with more than one tribe (three). The Asteraceae species from the Tageteae and Vernonieae tribes, each presented trophic associations with only one tephritid genus: *Trupanea* and *Tomoplagia*, respectively (**Table 2**).

Praxelis clematidea was the host in which the higher abundance of parasitoids (38), and presented trophic interaction with only one specimen of *Xanthaciura* sp.1. In the species of the tribes Eupatorieae and Heliantheae, ocurred higher florivore fly species richness, being obtained three species from each host plant. In the Heliantheae species were registered greatest abundance of Agromyzidae (136 adults), from which four of the six registered species of all the reared agromizids. In the species of Eupatorieae there was obtained higher abundance of tephritines (185), and 64 parasitoids (**Table 3**).

Characteristics of trophic interactions between florivorous flies and Asteraceae as a function of the Chaquean phytophysiognomies: Only the species *Trupanea* sp.1

Species of Asteraceae	Species of Tephritinae	Abundance
Tribe		
Eupatorieae		
Chromolaena ivifolia	*Cecidochares* sp.1	146
	Xanthaciura sp.2	20
Chromolaena margaritensis	*Xanthaciura* sp.1	17
Praxelis clematidea	*Xanthaciura* sp.1	1
Heliantheae		
Aspilia elata	*Dictyotrypeta* sp.1	23
	Dictyotrypeta sp.2	6
Dimerostemma grazielae	*Trupanea* sp.1	3
Tageteae		
Pectis odorata	*Trupanea* sp.1	17
Porophyllum angustissimum	*Trupanea* sp.2	16
Porophyllum ruderale	*Trupanea* sp.2	51
Vernonieae		
Lepidaploa remotiflora	*Tomoplagia minattai* Aczel 1955	2
Lessingianthus niederleinii	*Tomoplagia matzenbacheri* Prado, Norrbom & Lewinsohn, 2004	6

Table 3.
Species of Asteraceae infested by tephritine species (Tephritidae) and their absolute abundance in their flower heads sampled in the Brazilian Chaco (May 5, 2017 to April 5, 2018).

Florivorous Flies	Wooded Steppe Savanna	Park Steppe Savanna	Grassy-woody Steppe Savanna
Cecidochares sp. 1	0	0	146
Dictyotrypeta sp.1	0	23	0
Dictyotrypeta sp.2	0	6	0
Tomoplagia minattai	0	0	2
Tomoplagia matzenbacheri	0	0	6
Trupanea sp.1	3	16	1
Trupanea sp.2	31	33	3
Xanthaciura sp.1	8	7	3
Xanthaciura sp.2	0	0	20
Melanagromyza spp. (6 spp.)	3	81	87
Totals	45	166	268
Parasitoid Species Richness (S)	4	6	8
	19	50	27

Table 4.
Abundance of endophagous insects associated with Asteraceae chapters in three phytophysiognomies in the Brazilian Chaco (Porto Murtinho-MS) (May 5, 2017 to April 5, 2018).

and *Xanthaciura* sp.1 occurred simultaneously in the three phytophysiognomies. The species of *Dictyotrypeta* spp. were specific to PSS and *Cecidochares* sp.1 from GWSS. The two species of *Tomoplagia* recorded in this study, *Tomoplagia matzenbacheri* and *Tomoplagia minattai*, were specific to GWSS, as well as *Xanthaciura* sp.2.

None of the species of tephritids reported for WSS was specific to this phytophysiognomy (**Table 4**).

The phytophysiognomy with highest species richness (S = 6), and abundance of Tephritinae (268) was GWSS. The highest abundance of parasitoids (50) was found in the phytophysionomy Park Steppe Savanna (PSS). PSS presented the highest diversity by the Shannon (H) index, despite it haven't the greatest species richness (S). Probably this is due to the fact that this index takes into account the homogeneity of species in the environment. In the Grassy-Wood Steppe Savana (GWSS) the highest value of diversity was registered for the Margalef index. This index consider the sample size, which can be explained by the high abundance of the florivore fly *Cecidochares* sp.1 in this phytophysiognomy (**Tables 4** and **5**).

Applying the rarefaction method in the samples ("curve of collector"), it was possible to obtain estimates of the species richness of Tephritinae from the Chaquenha community. It indicated that the samples were insufficient to reach the asymptote of curve of collector. This means that the sampling effort was not enough to detect all species of florivorous flies present in the Brazilian Chaco. Due to the mosaic of phytophysiognomies in this biome, the few sampling points proved to be insufficient, to estimate the total species richness of florivorous flies in the Brazilian Chaco, even doing repetitions in the four year seasons (**Figure 1**).

Phytophysiognomies	H	ALFA	E	S
Wooded Steppe Savanna (WSS)	0.7285	0.5351	0.6631	3
Park Steppe Savanna (PSS)	1.4163	0.9077	0.88	5
Grassy-Woody Savana (GWS)	1.3084	0.9077	0.7302	7
Three combined Phytophysionomies	1.6927	1.4063	0.7302	9

Table 5.
Indices of Shannon (H), Margalef (A), equitability (E) and species richness (S) for Tephritinae (Diptera: Tephritidae) associated with flower heads of Asteraceae in three phytophysionomies from the Brazilian Chaco (Porto Murtinho -MS) (May 5, 2017 to April 5, 2018).

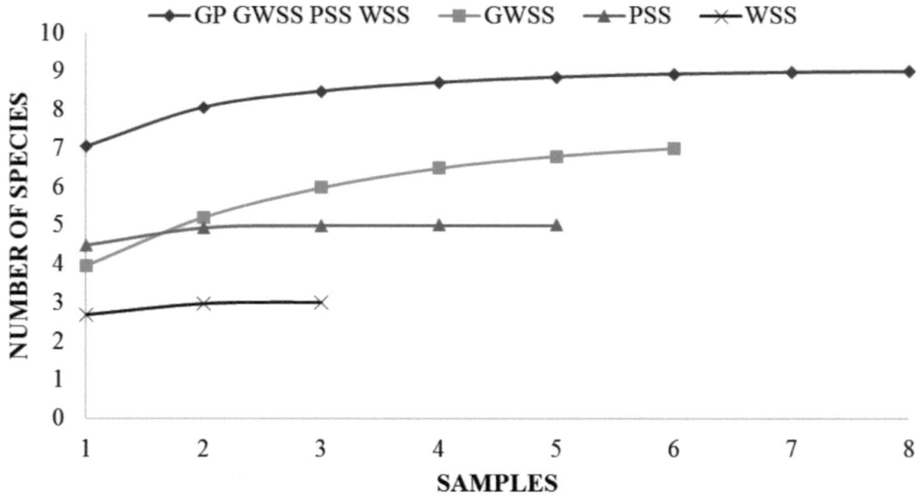

Figure 1.
Curve of species accumulation of Tephritinae (Diptera: Tephritidae) in three phytophysiognomies: General Panorama (GP), Grassy-Woody Steppe Savanna (GWS), Park Speppe Savanna (PSS), and Wood Speppe Savanna (WSS), in the Brazilian Chaco (Porto Murtinho-MS) (May 5, 2017 to April 5, 2018).

5. Discussion

The Brazilian Chaco presented a lower florivorous species richness than other neighboring Neotropical biomes already evaluated. Eighteen genera of Tephritinae (Tephritidae) and 80 species occurring in Brazil were reported by Prado [9]. In the Cerrado Biome, 12 genera are listed [10], and 10 genera have already been cataloged in the Atlantic Forest [11].

This research represents the first inventory of florivorous flies (Tephritidae and Agromyzidae) feeding in Asteraceae flower heads in the Chaco. Associations of *Melanagromyza* species (Agromyzidae) with Asteraceae species are reported for the Cerrado Biome [10], but there was yet no data for the Brazilian Chaco.

Due to the lack of keys for several taxa of Neotropical Tephritinae, many publications were unable to perform specific identification of florivorous flies. Species of *Cecidochares*, *Dicyotrypeta*, *Trupanea* and *Xanthaciura* are constantly only morphospecified.

In the Brazilian Chaco, the abundance of tephritids was higher than that of agromyzids, as well as the species richness (9 Tephritinea spp. against 6 probable species of *Melanagromyza* spp. (Agromyzinae, Agromyzidae). This pattern was also repeated in the Cerrado Biome [10]. According to several authors in the Neotropical Region, the most frequent and abundant species of florivore Diptera are the tephritines (Tephritidae). Herein, tephritines and agromyzines presented similar frequencies. Therefore, this pattern is not well understood and needs further studies in the Chaco biome to clarify its patterns of co-occurrence.

The Brazilian Chaco presented 16 trophic interactions between 15 species of florivorous flies and their Asteraceae species. In the Cerrado Biome, 49 species of Asteraceae are reported as hosts of florivore dipterans [10]. Herein, *Chromolaena margaritensis* hosted *Xanthaciura* sp.1 and *Melanagromyza* sp. 6. These are the first records for such associations worldwide. The flower heads of *C. margaritensis* has already been reported as a host for *Cecidochares* species in South Brazil [12].

Tomoplagia species were specific to the Vernonieae tribe. This relationship is already known [9]. In this research, *Trupanea jonesi* was the most generalist species, infesting two distinct asterace tribes: Heliantheae and Tageteae. The polyphagy of *T. jonesi* is well known, more than 100 associations with their host plants have been established [13].

The low diversity of florivorous flies recorded in the Brazilian Chaco can be explained by the low diversity of Asteraceae species there or by insufficient sampling effort. Probably, the species richness of florivorous flies is positively correlated with the species richness of sampled Asteraceae in the Chaco. Biomes richer in Asteraceae show a greater diversity of these trophic interactions. The Cerrado presents a high diversity of Asteraceae and associated species of Tephritinae, as has pointed out by other inventories [9, 11, 12].

Tomoplagia matzenbacheri and *Tomoplagia minattai* found in this research are new records for the state of Mato Groso do Sul (MS). The only previous work developed in the state of MS with florivorous flies was carried out by Uchoa, Wachter-Serapião & Roque [14] in the Cerrado, a fragment of the Atlantic Forest and an agroecosystem (orchard).

The Brazilian Chaco presented 25 species of Asteraceae, which apparently represents 92% of the species cataloged for this Biome [15]. However, many of the species recorded in this study do not appear in the floristic inventories of the Brazilian Chaco and vice-versa. New floristic studies must answer if there is a subsampling or if it is the result of the divergent interpretation among different authors about the characterization of phytophysiognomies truly Chaqueans.

Herein, *Dimerostemma grazielae* was host to florivorous flies in the Brazilian Chaco. This species of Asteraceae is considered rare in Brazil [16]. *Calea rupicola* that was recorded on the APA trail (GWS) (=APA Waterfall Municipal Park = *Parque Municipal Cachoeira do Apa*) is considered endemic to the state of Mato Grosso do Sul, also reported in the Pantanal region [15]. *C. rupicola* has not been associated with florivorous flies in this research. In the state of Mato Grosso do Sul there are, at least, 32 endemic species of Asteraceae.

The Wood Steppe Savanna (WSS) presented pioneer plants such as *Conyza bonariensis* and *Erechtites hieracifolius*, indicating that environment as the most degraded. A conjecture of elements is related to the low diversity in this place: deconfiguration of the flora by human action, reducing the species richness of Asteraceae, and as well as, some punctual flooding during the summer period, caused by anthropic action.

In this study, the occurrence of *Pectis odorata* was reported in the Brazilian Chaco. This plant was common in flooded areas, mainly in the WSS. *Pectis gardneri*, presents adaptations for the common water deficit in the Brazilian Chaco, as pointed out by Antunes [17]. New floristic studies can answer if there are also adaptations to constant flooding, in species of Asteraceae and other taxa in the Brazilian Chaco.

Herein, we added new data about biodiversity of Tephritinae (Tephritidae) and Agromyzinae (Agromyzidae), presenting essential information to fight for environmental preservation, as well as contribute to the catalog of flora and fauna from Chaco. Human-caused phenomena, such as global warming and habitat destruction, have increasingly threatened the planet's biodiversity. Endemic species are the most susceptible to disappear, due to their ecological sensitivity. As there is a dependence on florivorous flies for their host Asteraceae, a relationship extremely species-specific, both taxa have a greater chance of co-extinction. The savannas are at high risk of species extinction, due to the loss of habitat being much greater than the conservation efforts by human community and political authorities.

Finally, in the Brazilian Chaco: *Cecidochares* sp.1 was the most abundant species of florivorous fly upon the Asteraceae flower heads. A species of this same genus, *Cecidochares connexa* (Tephritinae) has been successful employed for the biological control of an exotic Asteraceae in Ghana, West Africa (*Chromolaena odorata*). This plant species was introduced on Africa by decade of 1930 and to humid regions of tropical Asia, with negative impacts on agriculture and regional biodiversity [4].

Trupanea species were the most polyphagous in the Chaco. From the Heliantheae tribe of Asteraceae we recovered the highest abundance of *Melanagromyza* spp. (Agromyzinae, Agromyzidae). From the species Eupatorieae tribe were reared highest abundances of tephritids and their parasitoids (Hymenoptera). Park Steppe Savanna (PSS) was the phytophysiognomy among the three evaluated that presented the highest diversity (H′) of Tephritinae, probably, due to the heterogeneity of this phytophysiognomies in the Brazilian Chaco.

6. Conclusions and perspectives

1. In the Brazilian Chaco do occur at least 15 florivore fly species, nine of Tephritinae (Tephritidae), from five different genera, and six species of *Melanagromyza* (Agromyzinae, Agromyzidae);

2. The Grassy-Woody Savanna is the Chaquean phytophysiognomy that harbored higher species richness;

3. Some florivore fly (Tephritinae and Agromyzinae) species needs to better studied to employ in programs of biological control for invasive Asteraceae in the Neotropical Region.

Acknowledgements

We thank the *Conselho Nacional de Desenvolvimento Científico e Tecnológio* (CNPq) by a research grant (Process No. 305112/2012–0) based on research productivity awarded to M.A. Uchoa, *Coordenação de Aperfeiçoamento de Pessoal de Nível Superior* (CAPES), for the MSc grant to Anderson Fernandes da Silva; *Fundação de Apoio ao Desenvolvimento do Ensino, Ciência e Tecnologia do Estado de Mato Grosso do Sul* (FUNDECT) & CAPES-*Edital Chamada N°. 44/2014 PAPOS-MS*); FUNDECT & CAPES-*Edital Chamada N°.* 12/2015-*BIOTA-MS: Ciência e Biodiversidade*, and Universidade Federal da Grande Dourados (UFGD), for the financial support to the Project Insects Biodiversity on the Brazilian Chaco.

Conflict of interest

The authors declare no conflict of interest.

Recommendations

More sampling points are needed to represent the real diversity of florivorous flies in the Brazilian Chaco, as well as, to quantify and qualify the endophagous insect interactions with Asteraceae on the Chaco Biome (Argentina, Bolivia and Paraguay).

To have a better understanding and a refined quantifications and qualification on the relationships between florivorous fly species, their natural enemies and Asteraceae, is important individualize capitulum samples to obtain the infesting insects or their respective parasitoids.

Author details

Manoel A. Uchoa[1*], Anderson S. Fernandes[1] and Jimi N. Nakajima[2]

1 Laboratory of Taxonomy and Systematics of Tephritidae (LabTaxon), Universidade da Grande Dourados (UFGD), Dourados, MS, Brazil

2 Universidade Federal de Uberlândia, Uberlândia, MG, Brazil

*Address all correspondence to: uchoa.manoel@gmail.com

References

[1] Lewinsohn TM (1991) Insects in flower heads of Asteraceae in southeast Brazil: a case study of tropical species richness. **Plant-animal interactions: evolutionary ecology in tropical and temperate regions**, pp.525-559.

[2] Pinto-Junior HV, Villa PM, Pereira MCA, Menezes LFT (2021) The pattern of high plant diversity of Neotropical inselbergs: highlighting endemic, threatened and unique species. Acta Botanica Brasilica. https://www.scielo.br/pdf/abb/2021nahead/0102-3306-abb-0102-33062020abb0129.pdf <Accessed on February 1, 2021>. Doi: 10.1590/0102-33062020abb0129

[3] Egli D, Olckers T (2017) Establishment and impact of insect agents deployed for the biological control of invasive Asteraceae: prospects for the control of *Senecio madagascariensis*. BioControl 62:681-692.

[4] Aigbedion-Atalor PO (2020) Weed or not a weed? Density, perceptions and management of *Chromolaena odorata* (Asteraceae) in West Africa: Voices from Ghana. Weed Research 60 (6): 406-414. Doi: 10.1111/wre.12439

[5] Prado DE (1993) What is the Gran Chaco vegetation in South America? Candollea 145:29

[6] Spichiger R, Calenge C, Bise B (2004) Geographical zonation in the Neotropics of tree species characteristic of the Paraguay-Paraná Basin. Journal of Biogeography 31: 1489-1501

[7] Uchoa MA (2012) Fruit Flies (Diptera: Tephritoidea): Biology, host plants, natural enemies, and the implications to their natural control, pp. 271-300. *In*: Larramendy, M. L. & Soloneski, S. (eds.). **Integrated Pest Management and Pest Control - Current and Future Tactics.** ISBN 978-953-307-926-4. DOI 10.5772/31613. InTechOpen. Rijeka, Croatia. 668p. https://www.intechopen.com/books/integrated-pest-management-and-pest control-current-and-future-tactics/fruit-flies-diptera-tephritoidea-biology-host-plants-natural-enemies-and-the-implications-to-their-n (Accessed on January 05, 2021).

[8] Fernandes AS, Uchoa MA (2019) Diversidade de moscas florívoras (Diptera: Agromyzidae e Tephritidae) e seus Parasitoides (Hymenoptera) em capítulos florais de Asteraceae no Chaco Brasileiro. Relatório para Exame de Qualificação. Programa de Pós-Graduação em Entomologia e Conservação da Biodiversidade, Universidade Federal da Grande Dourados (UFGD), Dourados-MS, Brazil. 58p.

[9] Prado PI, Lewinsohn TM, Almeida AM, Norrbom AL, Buys BD, Macedo AC, Lopes MB (2002) The Fauna of Tephritidae (Diptera) from Capitula of Asteraceae in Brazil. Proc Entomol Soc Washingt 104:1001-1028.

[10] Almeida AM, Fonseca CR, Prado PI, Almeida-Neto M, Diniz S, Kubota U, Braun MR, Raimundo RLG, Anjos LA, Tehra GM, Futada SM, Lewinsohn TM (2006) Assemblages of endophagous insects on Asteraceae in São Paulo cerrados. Neotrop Entomol 35:458-468.

[11] Savaris M, Lampert S, Lorini LM, Pereira PR, Marinoni L (2015) Interaction Between Tephritidae (Insecta, Diptera) and plants of the family Asteraceae: New host and distribution records for the state of Rio Grande do Sul, Brazil. Rev Bra Entomol 59:14-20.

[12] Garcia FRM, Sabedot-Bordin SM, Bogus GM, Bampi D (2011) Tefritídeos endófagos (Diptera: Tephritidae) associados à Asteraceae em Chapecó, Santa Catarina. Biotemas 24:15-20.

[13] Headrick DH, Goeden RD (1998) The Biology of Nonfrugivorous Tephritid Fruit Flies. Annu Rev Entomol 43:217-241.

[14] Uchoa, Wachter-Serapião & Roque (2021). Feeding by Florivorous Flies (Tephritidae and Agromyzidae) in Flower Heads of Neotropical Asteraceae (Asterales) from Central Brazil. https://www.intechopen.com/online-first/feeding-by-florivorous-flies-tephritidae-and-agromyzidae-in-flower-heads-of-neotropical-asteraceae-a

[15] Roque N, Teles AM, Moura L, Pacheco RA, da Silva GHL, Alves M, Nakajima JN (2018) Checklist of asteraceae in Mato Grosso do Sul state, Brazil. Iheringia - Ser Bot 73:147-156.

[16] Giulietti AM, Rapini A, Andrade MJG de, Queiroz LP, Silva JMC (2009) Plantas Raras do Brasil, Conservação Internacional, Belo Horizonte, 495p.

[17] Antunes AF. Adaptações estruturais de três espécies de Asteraceae do Chaco Brasileiro (2015). Dissertação de Mestrado. Universidade Federal de Mato Grosso do Sul. Campo Grande-MS, 49p.

Chapter 5

Feeding by Florivorous Flies (Tephritidae and Agromyzidae) in Flower Heads of Neotropical Asteraceae (Asterales) from Central Brazil

Manoel A. Uchoa, Morgana F. Wachter-Serapião and Nádia Roque

Abstract

The four following Diptera families are peculiar because they are predominantly phytophagous: Cecidomyiidae, Agromyzidae, Lonchaeidae and Tephritidae; which is uncommon for dipterans. Tephritine's larvae, depending on the species, consumes leaves, stems, flowers or roots of their host plants. Some tephritines feeds on flower heads of weed Asteraceae and can act in population suppression of invasive species in cultivated areas. In Mid-West of Brazil, we investigate Tephritinae and Agromyzinae flies in flower heads of Asteraceae species in three different phytophisiognomies in Dourados region, state of Mato Grosso do Sul. Here, 12 florivore fly species (9 Tephritinae, and 3 *Melanagromyza* spp., Agromizinae, Agromyzidae) are reported for the first time in Mid-West Brazil. We stablish the species of Asteraceae host for Tephritinae (Tephritidae) and for some species of *Melanagromyza* (Agromyzinae) in environments of Cerrado, Semideciduous Forest, and agroecosystem at Dourados-MS region. The inflorescences of Asteraceae species (± 500 capitula/species) were kept in containers to the emergence of the florivorous flies or their parasitoids. The adult insects after 48 hours were fixed in 80% ethanol for later identification. A total 36 species of Asteraceae were evaluated in the three regions of Dourados-MS, Brazil. Were obtained 120,031 flower heads of Astereceae, emerging 2,698 adults of insects: 833 Tephritinae (Tephritidae), belonging to 7 genera and 9 species; 1,089 *Melanagromyza* spp. (Agromyzidae) and 776 parasitoids (Hymenoptera) from the tephritines and agromyzines. We found that some florivore fly species needs to be better studied to employ in suppression programs of invasive Asteraceae population in the Neotropical Region.

Keywords: florivory, *Melanagromyza* spp., compositae, tephritinae, weed biocontrol

1. Introduction

Diptera is the second Order on diversity in the Class Insecta, but few families are predominantly phytophagous (feeds on plants), such as the ancient Cecidomyiidae, with almost 5,400 species worldwide (Bibionomorpha, Sciaroidea), and the three derived families of Brachycera: Agromyzidae (near 3,000 spp.) (Opomyzoidea), Lonchaeidae [around 600 spp.], and Tephritidae (about 5,000 spp.) (Tephritoidea). From the flies that feed on plants, these are the four families with higher diversity, totaling around 14,000 species.

The Cecidomyiids make galls in plants (like tumors) and feed inside these tissue that grows around their larvae. The three-following family of flies, are of higher economic importance: Agromyzids feed inside different parts of plants, being several species leaf or stem miners, with many pests (*e.g. Liriomyza* spp.) on cultivated vegetables and other crop plants. However, the larvae of species in the genus *Melanagromyza* Hendel 1920 feeds on the seeds of Asteraceae flower heads.

The Agromyzidae are represented by tiny phytophagous flies. Many species are known as leaf miners, with several species characterized as pests of some agricultural crops. Currently, there are almost 3,000 species described worldwide, being about 90 species reported in Brazil. Agromizids are generally species-specific in their host plants. They usually live and breed in a single plant genus, or at most in a family, except for a few pest species such as those of the genus *Liriomyza* which are very polyphagous.

Tephritidae is the most speciose family of fruit flies, with around 5,000 described species, in six subfamilies (Tachiniscinae, Blepharoneurinae, Phytalmyiinae, Trypetinae, Dacinae, and Tephritinae); about 500 genera, and probably many undescribed species worldwide. Tephritids are peculiars because they are among the few groups of dipterans strictly phytophagous, except the Tachiniscinae, which are thought be parasitoids of Lepidoptera, and at least, some species of Phytalmyiinae feed on live or dead bamboos (Poaceae), on fallen or dead trees of other plant families. Blepharoneurinae feed in flowers, fruits, and make galls in Cucurbitaceae; Trypetinae and Dacinae feed in fruits or in seeds of a wide range of plant families, and Tephritinae eat in flowers, make gall, or are leaf-miners in several plant families, such as: Aquifoliaceae, Scrophulariaceae, Verbenaceae, but mainly in flower heads of Asteraceae. Tephritinae is the biggest subfamily of Tephritidae, with around 1,840 described species (valid names) in 11 tribes and 211 genera [1–3].

Some tephritines (Tephritidae) genera are stems gall makers, such as *Eurosta* Loew 1873, *Procecidochares* Hendel 1914, the Neotropical *Tomoplagia* Coquillett 1910; others are florivorous like the species of Neotropical genus *Blepharoneura* Loew 183, that breed specifically in Cucurbitaceae's flowers or fruits. Many other genera (*e.g. Dictyotrypeta* Hendel 1914, *Dioxyna* Hardy 1988, *Cecidochares* Bezzi 1910, *Tetreuaresta* Hendel 1928, *Trupanea* Schrank 1795, *Xanthaciura* Hendel 1914), feed and breed in flower heads of daisis (Asteraceae). In the Neotropical Region, several tribes of Tephritinae and some species of *Melanagromyza* Hendel 1920 (Agromyzinae, Agromyzidae), breed manly upon the flower heads of Asteraceae - the most speciose plant family worldwide.

Florivorous flies are here defined as the dipterans that in their larval phase feed in the flowers of Angiospermae. The larvae of these flies, like other higher Brachycera, undergo through three instars (L1, L2 and L3), generally, doubling in size and weight after each molting of exoskeleton. After completing the third instar, larva expand their exoskeleton, assuming a barrel shape, being the pupa formed inside this last skin. These set (L3 exoskeleton + skin of the pupae) is called puparium. When the pupa inside the puparium is completely formed, that take a few days (around 2 weeks), the inner fly makes a circular hole in the cephalic pole of puparium, and emerges as an adult. Pupation of florivorous flies (Tephritinae and Agromyzinae) on the Neotropical Region, generally, happens inside the flower heads of Asteraceae.

The true flies differ from other groups of insects in this aspect: the skin of their larval body (exuviae) at the end of the third (last) instar is not lost, becoming expanded and hardened, and the pupa is formed internally, having in these phase, two protections. The beginning of this stage is recognized as the **pre-pupa phase**. Like in all other holometabolous insects, when the larva has completed the last instar, starts the phase of pre-pupae, which is characterized by the ending of juvenile feeding activity, expelling all hindgut feces, and starting the organogenesis to build up the pupa.

The adults of Tephritidae, differently of their larvae, which have a restrict kind of food (tissue of the host plant), they feed in different materials found in their environments. Probably they drink fruit juices, extrafloral exudates, nectar, honeydew produced by ancient Hemiptera, microorganisms and other sources of carbohydrates and aminoacids, such as exudates of plant (alive or dead), bird faces, etc. By other hand, Norrbom [1] has pointed out that adults of some Blepharoneurinae species, have labellar teeth in their mouth, being able to rasp plant tissue.

Asteraceae (Asterales) is the most specious family of Angiospermae, with around 24,000 species worldwide [4], having a great diversity of species in Brazil. These plants are adapted to stressful water regimes, hence their great capacity to colonize the most different ecosystems.

Female of florivorous flies (Tephritinae and Agromyzinae) after mate, lay their eggs over the flower buds or inside the young flowers of host plants. After 3–7 days the eggs hatch and larvae start eating, mainly the seeds, to complete their larval phase, emerging from the inflorescences as adult.

2. Protocol to evaluation the diversity of florivorous flies in nature

The sample the diversity of florivorous flies (FF) and their host plants, one can start collecting flower heads of Asteraceae. The inflorescences (capitula) can be held in small containers with same substrate that can keep humidity, such as toilet paper, sterile sand or vermiculite. Depending on the proposal of research, it is possible put the flower heads (around 500) in a container or individually (a flower head or a capitula in each container).

To sample the pattern of diversity for both: Host plants and FF is possible to employ different methodologies. For example, it's possible to collect samples of Asteraceae flower heads by transect; quadrants; in different sub-environments (Phytophysiognomies); using points determined by GPS, and so on. But some procedures are essential: All the plant from which flower heads were sampled must be exsiccated and identified by botanist specialist to have accurate information. The FF needs to be held alive for a couple of days (2 or 3), to allow that adult can acquire specie-specific coloring patterns of body and wings (important to specific and secure species identification).

Below we will present a characterization of some Tephritinae sampled in our recent researches, a synopsis on the main results on the Neotropical FF, their host plants and natural enemies (parasitoids) in natural environments and agroecosystem.

3. Characterization of sampled neotropical Tephritinae

Trupanea Schrank 1795 (**Figure 1a**), with 226 described species, is the largest genus of Tephritinae. It occurs in all biogeographic regions. Some Neotropical species have been reviewed and included in keys by Hendel, Malloch, Hering, Aczel, Frias, and Foote, but all are obsolete by now. The genus *Trupanea* is in need of review [1, 2].

Figure 1.
Some Neotropical genera of Tephritinae (Diptera: Tephritidae) breeding in flower heads of Asteraceae species from Central Brazil: a): Male of Trupanea *Schrank 1795; b): Female of* Tomoplagia *Coquillett 1910; c): Female of* Dioxyna *hardy 1988; d): Female of* Xanthaciura *Hendel 1914; e): Male of* Cecidochares *Bezzi 1910; f): Female of* Dictyotrypeta *Hendel 1914; b): Male of* Tetreuaresta *Hendel 1928.*

The genus *Tomoplagia* Coquillett 1910 (**Figure 1b**), with 61 described spcies and several one not described. From this total, at least 15 species are reported in the Central American region (including Neotropical Mexico). *Tomoplagia* species have interactions with Asteraceae, mainly in the genera of Vernonieae and Mutiseae, Heliantheae and other tribes. Most species reproduce in flower heads, but a South American species forms galls [1–3].

The genus *Dioxyna* Hardy 1988 (**Figure 1c**), has a wide distribution, and 11 described species. In the New World their hosts plants are mainly species of the tribes Heliantheae and Heleniae [1, 2].

The genus *Xanthaciura* Hendel 1914 (**Figure 1d**), includes 17 described species, and several not described ones. At least 10 occur in the Central American region (including Neotropical Mexico). Most species were reviewed by by Aczel.

Xanthaciura has species that consume flowers of several genera of Asteraceae, mainly in the tribes Eupatorieae, Coreopsideae and Heliantheae [1, 2].

The genus *Cecidochares* Bezzi 1910 (**Figure 1e**), includes 13 described and numerous not yet undescribeds. At least three undescribed species occur in the Central American region (and Neotropical Mexico). Species of *Cecidochares* are mainly related to plants of the Eupatorieae tribe (Asteraceae). They make galls on stems, and occur with low incidence on flower heads. *Cecidochares connexa* has been introduced in several Old World countries for the biological control of the important weed: *Chromolaena odorata* [1, 2].

Dictyotrypeta Hendel 1914 (**Figure 1f**) is a Neotropical genus in need of revision. Currently it includes six species and many undescribed ones. It is not clear if all *Dictyotrypeta* species form a monophyletic group. Several of them, such as *Dictyotrypeta incisa* (Wulp) from Central America, South America, Mexico, Guatemala, and *Dictyotrypeta crenulata* (Wulp) from Mexico (Sinaloa, Guerrero, Veracruz), were only tentatively included in this genus. In addition, to the two species above, the fauna from Central America also includes two undescribed species. The known host plants for *Dictyotrypeta* spp. are mainly Asteraceae species in the tribes Heliantheae and Vernonieae [1, 2].

Tetreuaresta Hendel 1928 (**Figure 1g**), a Neotropical genus with 19 species described, six of which occur in the Central American Region (including Neotropical Mexico), and numerous undescribed species. Steyskal provided a key for 5 species, but the genus is in need of revision. The biology of most species is unknown, but five species consume flower heads of Vernonieae (Asteraceae) species. *Tetreuaresta obscuriventris* (Loew) has been employed for biological control of *Elephantopus mollis* in Hawaii, and other Pacific Islands [1, 2].

4. Aims

The main objectives of this paper, are: 1.To investigate the occurrence of Florivorous fly (FF) species associated with Asteraceae flower heads in three phytophysiognomies from Dourados Region; 2. To verify which Asteraceae species are hosts of Tephritidae and Agromyzidae (Diptera) in Cerrado, Semideciduous Forest and Agroecosystem; 3. To quantify the patterns of occurrence of those dipterans and their parasitoids.

5. Material and methods

The work was developed in a tropical region, marked by a landscape dominated by Cerrado Biome, with patches of Semideciduous Forest (Atlantic Forest), and a matrix area of monocultures. The area has three biodiversity hotspots (Cerrado, Atlantic Forest and Chaco). The Cerrado that represents the phytophysiognomy of Savana Parque is located in the district of Itahum (22° 05 ‘21.8” S and 55°21′11.9” W), Dourados-MS, with 412 m of altitude. The second environment is a Seasonal Semideciduous Forest, belonging to the Atlantic Forest biome and represented by a fragment of native vegetation, with about 35 ha (22° 12 ‘46.7” S and 54° 54 ‘53.5” W) and altitude of 452 m. The third environment is a mixed fruit orchard at the *Universidade Federal da Grande Dourados* (UFGD), located at km 12 of the MS-162 Highway (22° 11′ 46.8” S, and 54° 56′ 12.2” W), at 425 m altitude, in the municipality of Dourados-MS.

Asteraceae flower heads were collected and kept in two transparent plastic cups (500 ml) with the openings juxtaposed and secured by adhesive stick tape.

The bottom cup had a layer of dry sterile sand to absorb excess moisture to allow the emergence of flies and/or their parasitoids, since the identification of species of tephritids is based on the morphology of adults. The sampled material was kept in the laboratory under a 12 h photophase, controlled by a timer.

After emergence, adults were fed an artificial diet composed of brown sugar (100 g), 50 g of beer yeast, a tablespoon of honey from *Apis mellifera* L., and 100 ml of sterile water. All adults were kept alive for at least 48 hours for complete pigmentation of the body and wings. After this period, they were killed and kept in bottles with 80% ethanol for specific identification and analysis of qualitative and quantitative data. In this research, 1 ha quadrant was established as the sample area in each evaluated environment. Sampling was repeated every 15 days, and collections were performed on the same day in the three different environments. Tephritids and agromyzids were identified in the *Laboratório de Sistemática e Taxonomia de Tephritidae* (LabTaxon)-UFGD.

The collected plants were processed for species identification by Dr. Nádia Roque, *Universidade Federal da Bahia* (Salvador, BA, Brazil). All exsiccates were coded for each individual and location, and duplicates of these were deposited as voucher specimens at the CGMS Herbarium of the *Museu da Biodiversidade, Faculdade de Ciências Biológicas e Ambientais (FCBA), Universidade Federal da Grande Dourados* (UFGD), Dourados-MS, Brazil, and at the Alexandre Leal Costa Herbário (ALCB) Institute of Biology, Universidade Federal da Bahia (UFBA, Salvador, Brazil). The level of infestation of each host plant was calculated according to the number of flower heads and the biomass of each plant species.

The analyzes were made by analysis of variance (ANOVA and F test), when the data meet the assumptions of normality and homogeneity, using, in this case, the means comparison tests (Tukey, Duncam or t test) at 10% of significance. When the assumptions were not met, Kruskall-Wallis and Mann–Whitney methods were used, respectively. The level of infestation of each host plant was calculated according to the number of flower heads and the biomass of each species of plant. Species richness, abundance, and frequency among the three different environments were evaluated.

5.1 Cerrado biome

Cerrado is an International Biome, important hotspot on biodiversity. Is present from Central Brazil, Bolivia, extending to Southern to north of Amazonia; across the isthmus of Central America, including the Caribbean region; Belize, Guatemala, Southern Mexico, and north of the Mexican Plateau. The Cerrado is the second largest Neotropical biome (behind Amazon = Tropical Andes), and fight an important hotspot of biodiversity worldwide. The Cerrado biome is located mainly in Central Brazil, with approximately 2,000,000 km^2 (25% of Brazilian territory), with different Phytophysiognomies. Many vegetal formations are similar to savanna formations, but some are similar to tropical forests, and riparian forests. This biome has a great diversity of plants and animals, with high rates of plant endemisms, with around 2% of global diversity [5]. Nowadays, the Cerrado biome is recognized as the sixth hotspot of the planet, among 34 listed in order of priority for conservation of habitat [6].

5.2 Atlantic forest biome

The Atlantic Forest is characterized by on of the highest rates of biodiversity and endemism at Neotropical region end worldwide. It's an international biome restricted to three countries in South America: Brazil, Argentina and

Paraguay. It follows the coastal region of the Atlantic Ocean along Brazil and part of Argentina (Province of Misiones), and by land, a part to the southwest of Paraguay. It is characterized by an extensive forest that varies from humid to dry type (plants with broad leaves), tropical forests (desciduous, semi-desciduous and sub-montana), fields of tropical and subtropical pastures, arboreal savanna, steppe savanna and grassy-wood savanna and mangrove. Brazil is home, by far, to the largest stretch of the Atlantic Forest, with an extension of over 4,000 km, extending from the Rio Grande do Norte to the Rio Grande do Sul states. The Atlantic Forest is subdivided into ecoregions that are characterized by housing a huge biodiversity with high rates of endemism of animals and plants. Currently it is very devastated. Of the more than 1,315,000 km2 original from the time of the discovery of Brazil in the year 1,500 by the Portuguese colonizers, there are currently about 8.5% left, having already led several species of animals and plants to extinction, and with about 1,990 species at risk of extinction, according to IBGE [6].

5.2.1 Semideciduous Forest

It's a phytophysiognomy sub-type of the Atlantic Forest, characterized by plant species with senescence of their leaves, which are partially lost on the cold-dry seasons, like end of fall and start of the winter. Their leaf lost is dependent on the weather (coldness-dryness), but the tree renew their leaves/when the climate is mild dor worm or do not lost their leaves when/where the climate keeps constant.

6. Results and discussion

A total 36 species of Asteraceae (13 tribes) were sampled in three environments from regions of Dourados-MS, Mid-West of Brazil, looking for florivorous flies in their flower heads. Eighteen Asteraceae species hosted Florivorous flies (Tephritidae and/or Agromyzidae), and some of their parasitoids (Hymenoptera). From 18 sampled asteraceae species, neither Tephritidae nor Agromyzidae emerged. Twenty seven, species of Asteraceae were recorded in the Cerrado environment (Itahum District), 24 in the Semideciduous forest (phytophisiognomy of the Atlantic Forest, located at Fazenda Coqueiro), and 8 Asteraceae species were sampled in the Agroecosystem (a diversified orchard of fruit trees at UFGD campus) (**Table 1**).

The percentage of frequency of Tephritinae species associated with the species of Asteraceae were evaluated. In the Cerrado, we found higher diversity of Asteraceae (S = 27), being 11 of them host for florivorous flies (FF), emerging 12 species (9 Tephritinae and 3 Agromyzinae) from 374 adults recovered. In the Semideciduous forest, 24 species of Asteraceae (S = 24) were found. Ten of them were hosted by 9 florivorous fly species. Finally, in the agroecosystem, five Asteraceae species hosted four FF species (**Table 2**).

Eighteen Asteraceae species hosted the florivore fly species. Seven astereces had their flower heads coninized simultaneously by tephritines plus agromyzines: *Baccharis triplinervis*, *Bidens pilosa*, *Chaptalia integerrima*, *Chromolaena arnottiana*, *Porophyllum ruderale*, *Vernonia cognata*, and *Vernonia polyanthes*. Fom six host plants (*Chromolaena ivifolia*, *Eupatorium multicrenulatum*, *Praxelis pauciflora*, *Pterocaulon virgatum*, *Vernonia bardanoides* and *Zinnia elegans*) only species of tephritine emerged. Five asteraces were excluvive hosts for agromyzines of the genus *Melanagromyza* (*Aspilia latissima*, *Bidens sulphurea*, *Emilia fosbergii*, *Lourteigia ballotifolia*, and *Sonchus oleraceus* (**Table 2**).

Asteraceae Taxa (Tribes and Species)	Plant Status	Environment
Anthemideae *Tanacetum vulgare* L.	Nonhost	Agroecosystem (= Orchard)
Astereae *Baccharis linearifolia* (LAM.) Pers.	Nonhost	Cerrado Semideciduous Forest
Baccharis triplinervis (Spreng.)	Host	Cerrado
Conyza bonariensis (L.) Cronquist	Nonhost	Agroecosystem, Semideciduous Forest
Solidago microglossa DC.	Nonhost	Cerrado Semideciduous Forest
Cichorieae *Sonchus oleraceus* L.	Host	Cerrado Semideciduous Forest
Cynareae *Arctium lappa* L.	Nonhost	Semideciduous Forest
Eupatorieae *Chromolaena arnottiana* (Griseb.) R.M.King & H. Rob.	Host	Cerrado Semideciduous Forest
Chromolaena ivifolia (L.) R.M.King & H. Rob	Host	Cerrado
Eupatorium macrocephalum (Less.) DC.	Nonhost	Cerrado
Eupatorium multicrenulatum Sch. Bip. ex Baker	Host	Cerrado
Eupatorium odoratum (L.) King & H.E. Robins	Nonhost	Semideciduous Forest
Lourteigia ballotifoli (Kunth) R. M. King & H. Rob. N. V.	Host	Cerrado
Mikania hastato-cordata Malme	Nonhost	Cerrado Semideciduous Forest
Praxelis pauciflora (Kunth) R.M.King & H. Rob.	Host	Agroecosystem
Gnaphalieae *Achyrocline satureioides* (LAM.) D.C.	Nonhost	Cerrado
Heliantheae *Aspilia elata* Pilg.	Nonhost	Cerrado Semideciduous Forest
Aspilia latissima Malme	Host	Cerrado Semideciduous Forest
Bidens pilosa L.	Host	Agroecosystem, Semideciduous Forest
Bidens sulphurea (Cav.) Sch. Bip. N. V.	Host	Cerrado Semideciduous Forest
Salmea scandens (L.) DC.	Nonhost	Cerrado Semideciduous Forest
Tridax procumbens L.	Nonhost	Agroecosystem, Cerrado
Unxia kubitzkii H. Robinson	Nonhost	Semideciduous Forest
Zinnia elegans Jacq.	Host	Cerrado
Lactuceae *Hypochaeris brasiliensis* (Less.) Griseb.	Nonhost	Agroecosystem Cerrado Semideciduous Forest
Mutisieae *Chaptalia integerrima* (Vell.) Burkart	Host	Semideciduous Forest
Plucheae *Pterocaulon virgatum* (Lam.) DC.	Host	Semideciduous Forest

Asteraceae Taxa (Tribes and Species)	Plant Status	Environment
Senecioneae *Emilia fosbergii* Nicolson. N.V.	Host	Agroecosystem, Cerrado
Erechtites hieracifolia (L.) Raf.	Nonhost	Cerrado Semideciduous Forest
Targeteae *Porophyllum ruderale* (Jacq.) Cass.	Host	Agroecosystem, Cerrado Semideciduous Forest
Vernonieae *Cyrtocymura scorpioides* (Lam.) H. Rob.	Nonhost	Cerrado Semideciduous Forest
Vernonia bardanoides Less.	Host	Cerrado
Vernonia cognata Less.	Host	Cerrado Semideciduous Forest
Vernonia polyanthes Less.	Host	Cerrado Semideciduous Forest
Vernonanthura brasiliana (L.) H. Rob.	Nonhost	Cerrado Semideciduous Forest
Vernonanthura chamaedrys (Less.) H. Rob.	Nonhost	Cerrado Semideciduous Forest

Table 1.
Status for Asteraceae species to florivorous flies Tephritinae (Tephritidae, and/or Melanagromyza Hendel 1920, Agromyzinae: Agromyzidae), and environment of occurrence in Dourados region, MS, Brazil.

Nine Tephritinae species from seven genera were obtained: The recovered species were: *Trupanea jonesi*, *Tomoplagia brasiliensis*, *Tomoplagia reimoseri*, *Dioxyna chilensis*, *Xanthaciura unipuncta*, *Xanthaciura biocellata*, *Cecidochares fluminensis*, *Dictyotrypeta* sp. and *Tetreuaresta* sp. form 13 Asteraceae species [7]. From the flower heads of 11 species of plants were also obtained three species of *Melanagromyza* (Agromyzina, Agromizidae) from all three environments (**Table 2**).

In the total were obtained 120,031 flower heads of Astereceae, emerging 2,698 adults of insects: 833 Tephritinae (Tephritidae), belonging to 7 genera and 9 species; 1,089 *Melanagromyza* spp. (Agromyzidae) and 776 parasitoids (Hymenoptera) from the tephritines and agromizines. A total of 374 adults of Tephritinae were reared from the flower heads collected on the Cerrado, 269 from the Semideciduous Forest, and 190 fom the Agoecosystem. From the Agroecosystem seven asteraces were sampled (S = 7), 190 individuals of 4 species (2 Tephritinae / 2 Agromyzinae), emerged. In general, the Agromyzidae were more abundant than the Tephritidae (n = 1,089), but the Tephritinae were more biodiverse (nine species) than the Agromyzids, represented by three species. Some 776 adults of Hymenoptera parasitoids emerged from puparium of both families (Tephritidae and Agromyzidae) of florivorous flies (**Tables 1** and **3**).

The collected flower heads give a biomass of 8,202 grams, being 20,766 (5.7%) of the flower heads infested by FF, corresponding to a biomass of 1,587 g (5.16% of total). The species with the highest infestation rates for Tephritinae were: *Chaptalia integerrima*, with 0.500 fly by flower head (FH) and *Chromolaena ivifolia* with 8.09 fly/g. The lowest indexes occurred in *Sonchus oleraceus*, with 0.002 fly/FH and 0,009 individual/g, and *Pteurocaulom virgatum* with 0.010 fly/FH and 0.095 fly/g. For *Melanagromyza* species (Agromyzinae, Agromyzidae), *Bidens pilosa* was the Asteraceae with the highest rate of infestation: 0.079 fly/FH and 0.584 fly/g. The lowest indexes occurred in *Sonchus oleraceus*, with 0.002 fly/FH and 0.009 fly/g

Asteraceae Species	**Environments**		
	Cerrado	**Semideciduous Forest**	**Agroecosystem (= Mixed Orchard)**
Aspilia latissima	—	*Melanagromyza* sp.3	—
Baccharis triplinervis	*Xanthaciura unipuncta*	—	*Melanagromyza* sp.2
Bidens pilosa	*Melanagromyza* sp.1	*Dioxyna chilensis* *Melanagromyza* sp.1	*Dioxyna chilensis* *Melanagromyza* sp.1
Bidens sulphurea	*Melanagromyza* sp.2	—	—
Chaptalia integerrima	—	*Trupanea jonesi* *Melanagromyza* sp.2	—
Chromolaena arnottiana	*Trupanea jonesi* *Tomoplagia reimoseri* *Xanthaciura unipuncta* *Xanthaciura biocellata* *Cecidochares fluminensis* *Melanagromyza* sp.3	*Dioxyna chilensis* – – – –	– – – – –
Chromolaena ivifolia	*Trupanea jonesi* *Xanthaciura biocellata* *Cecidochares fluminensis*	– – –	– – –
Emilia fosbergii	-	—	*Melanagromyza* sp.2
Eupatorium multicrenulatum	—	*Xanthaciura biocellata*	—
Lourteigia ballotifolia	*Melanagromyza* sp.3		
Porophyllum ruderale	*Trupanea jonesi* *Dioxyna chilensis*	*Trupanea jonesi* *Dioxyna chilensis*	*Trupanea jonesi* *Dioxyna chilensis* *Melanagromyza* sp.1
Praxelis pauciflora	—	—	*Trupanea jonesi*
Sonchus oleraceus		*Melanagromyza* sp.3	
Pterocaulon virgatum	– –	*Xanthaciura biocellata* *Tetreuaresta* sp.	– –
Vernonia bardanoides	*Tomoplagia brasiliensis* *Tomoplagia reimoseri*	– –	– –
Vernonia cognata	*Xanthaciura biocellata* *Melanagromyza* sp.3	*Trupanea jonesi* *Xanthaciura biocellata* *Cecidochares fluminensis*	– – –
Vernonia polyanthes	*Tomoplagia brasiliensis* *Tomoplagia reimoseri* *Dioxyna chilensis* *Xanthaciura unipuncta* *Xanthaciura biocellata* *Dictyotrypeta* sp. *Melanagromyza* sp. 3	*Xanthaciura biocellata* *Tomoplagia reimoseri* *Cecidochares fluminensis* – – *Melanagromyza* sp.3	– – – – – –
Zinnia elegans	*Xanthaciura unipuncta* *Xanthaciura biocellata*	— —	— —

Table 2.
Florivorous fly species (Tephritidae and Agromyzidae) associated with flower heads of Asteraceae (Asterales) species in three environments from the region of Dourados-MS, Brazil (January 2011 to august 2012).

Tephritidae	Cerrado	Semideciduous Forest	Agroecosystem	Total
Trupanea jonesi	17	13	30	**60**
Tomoplagia brasiliensis	16			**16**
Tomoplagia reimoseri	16	5		**21**
Dioxyna chilensis	121	239	160	**520**
Xanthaciura unipuncta	70			**70**
Xanthaciura biocellata	130	8		**138**
Cecidochares fluminensis	3	2		**5**
Dictyotrypeta sp.	1			**1**
Tetreuaresta sp.		2		**2**
Subtotal Tephritidae	**374**	**269**	**190**	**833**
Agromyzidae				
Melanagromyza sp.1	73	767	173	**1,013**
Melanagromyza sp. 2	16	32	15	**63**
Melanagromyza sp. 3	10	3	0	**13**
Subtotal Agromyzidae	**99**	**802**	**188**	**1,089**
Parasitoids (Hymenoptera)	259	271	246	**776**
Subtotal	**732**	**1,073**	**632**	—
Total of Insects associated to Asteraceae species		**2,698**		

Table 3.
Abundance of Tephritidae and Agromyzidae (Diptera) sampled in the phytophysiognomies: Cerrado, Semideciduous Forest, and agroecosystem in the three subregions of Dourados-MS, Brazil (January 2011 to august 2012).

(**Table 4**). From *Aspilia latissima* only *Melanagromyza* sp.3 emerged, no tephritines were obtained. By other hand, both taxa of florivorous flies: tephritines and some unidentified species of *Melanagromyza* (Agromyzinae) shared the host species of Asteraceae, simultaneously (**Table 4**).

The association of Tephritinae species with Asteraceae species was compared using a symmetric normalization model, validated by chi-square. This association was highly significant {$x^2 = 492.72$; g.l = 288; ($p < 0.000$)}, explaining 54.8% of the total results. The Tephritidae correlated with the three phytophysiognomies, and with the Asteraceae species. The association of the species of florivorous fly with species of Asteraceae followed the model of symmetric normalization, also validated by chi-square. This relationship was highly significant {$x^2 = 93.407$; g.l = 16; ($p < 0.000$)}, explaining 100% of all results. Emerged some parasitoids from Asteraceae, being associated with the florivorous flies. All of them are Hymenoptera, and still wait for identification (**Table 5**).

There was a significant difference between the species *T. jonesi* with *X. biocellata*, *D. chilensis* and *X. unipuncta*, with a lower average of individuals [7]. *Xanthaciura biocellata*, *D. chilensis* and *X. unipuncta* were the most abundant Tephritinae species, resulting in a standard deviation with little variability in relation to the mean (**Table 5**).

The abundance of six avaluated species of florivorous flies in the Semideciduous Forest environment was significantly lower than in the agroecosystem, and the Cerrado environment did not differ from the other phytophysiognomies (**Table 6**).

ASTERACEAE SPECIES	Nunber of Flower heads	Biomass of Flower heads	Abundance: Tephritinae and Agromyzinae	*IL- FF/Flower heads	**IL-FF/g
TEPHRITINAE					
Baccharis triplinervis	400	13	7	0.017	0.538
Bidens pilosa	3,114	424	248	0.079	0.584
Chaptalia integerrima	18	6	9	0.500	1.500
Chromolaena arnottiana	1,750	48	114	0.065	2.375
Chromolaena ivifolia	500	11	89	0.178	8.090
Eupatorium multicrenulatum	1,000	11	2	0.002	0.181
Porophyllum ruderale	6,228	639	310	0.049	0.485
Praxelis pauciflora	100	30	3	0.003	0.100
Pterocaulom virgatum	186	21	2	0.010	0.095
Vernonia bardanoides	500	70	8	0.016	0,114
Vernonia cognata	2,000	25	9	0.004	0.36
Vernonia polyanthes	3,500	165	15	0.004	0.90
Zinnia elegans	70	107	11	0.157	0.102
AGROMIZYDAE					
Aspilia latíssima	61	108	2	0.032	0.0185
Baccharis triplinervis	2,567	168	13	0.005	0.026
Bidens pilosa	3,688	483	1,012	0.274	2.095
Bidens sulphurea	269	65	9	0.033	0.138
Chaptalia integerrima	180	60	14	0.077	0.233
Chromolaena arnottiana	500	10	1	0.002	0.100
Emilia fosbergii	611	98	2	0.003	0.020
Lourteigia ballotifolia	1,668	56	9	0.005	0.160
Porophyllum ruderale	186	21	1	0.005	0.047
Sonchus oleraceus	447	111	1	0.002	0.009
Vernonia cognata	500	4	2	0.004	0.500
Vernonia polyanthes	2,000	104	22	0.011	0.211

**IL- FF / Flower heads = Infestation Level Florivorous Flies by Asteraceae Flower Head.*
***IL-FF / g (mass) of Asteraceae Flower heads.*

Table 4.
Levels of infestation by florivorous flies (Diptera: Tephridae and Agromyzidae) in Asteraceae species sampled in three different ecosystems in the region of Dourados-MS, mid-west of Brazil (January 2011 to august 2012).

The frequency of the three species of the genus *Melanagromyza* (Agromyzidae) was correlated with their host plants. A significant frequency index of *Melanagromyza* spp. was found in *Bidens pilosa*. To the *Melanagromyza* spp. the "*picão-preto*" (a weed), *Bidens pilosa*, was the Asteraceae species with the highest abundance and frequency of *Melanagromyza* sp.1, representing 73.95% of the Agromyzidae in this Asteraceae species. In others Asteraceae species, the frequency of occurrence to *Melanagromyza* spp. was 7% or less (**Table 7**).

In Neotropical Region there are few reports for occurrence of tephritids and other florivorous flies in Asteraceae. There are some studies in order to inventory

Species de Tephritinae	n	Ȳ ± SD*
Trupanea jonesi	31	1.97[a] ± 2.01
Tomoplagia brasiliensis	5	2.20[abcd] ± 1.79
Tomoplagia reimoseri	9	2.3[abc] ± 2.23
Xanthaciura biocellata	22	6.27[be] ± 6.84
Xanthaciura unipuncta	14	5.36[ce] ± 4.98
Dioxyna chilensis	90	5.79[de] ± 5.74

**Kruskal Wallis Test = {P(x > X²); (α < 0.01)}. The Mean comparison by Mann–Whitney Test at 5%. Equal lower-case letters to vertical or column and upper-case letters equal to horizontal or line, do not differ significantly; n: number or sample size; Ȳ ± SD: Mean ± Standard Deviation*

Table 5.
Tephritinae (Diptera: Tephritidae) associated with the flower heads of Asteraceae (average and standard deviation) in the Dourados-MS region, mid-west of Brazil (January 2011 to august 2012).

Phytophysiognomies	n	Ȳ ± SD*
Semideciduous Forest	68	3.96[a] ± 4.54
Cerrado	67	4.67[ab] ± 5.27
Agro-ecosystem	44	5.75[bc] ± 6.28

**Average Comparison Test (Duncan = 10%), being letters equal to each other do not differ Significantly; n: number or sample size; Ȳ ± SD: Mean ± Standard Deviation*

Table 6.
Number of individuals of Tephritinae (Diptera: Tephritidae) obtained from Asteraceae flower heads in three phytophysiognomies [comparison of means and standard deviation] in Dourados region-MS, mid-west of Brazil (January 2011 to august 2012).

the number of species associated with the flower heads of Asteraceae in South, and Southeast of the Brazil.

Our recent researches in the Brazilian Mid-West have established the fowling relationships among florivorous fly species and Asteraceae flower heads: *Dioxyna chilensis* reared from *Bidens pilosa*; *Tomoplagia reimoseri* of *Chromolaena arnottiana*, *Vernonia bardanoides* and *Vernonia polyanthes*; *Trupanea jonesi* of *Chaptalia integerrima*, *Chromolaena arnottiana*, *Chromolaena ivifolia*, *Porophyllum ruderale*, *Praxelis pauciflora* and *Vernonia cognata*; *Cecidochares fluminensis* in *Chromolaena arnottiana*, *Chromolaena ivifolia*, *Vernonia cognata* and *Vernonia polyanthes*, *Dictyotrypeta* sp. in *Vernonia polyanthes* [7].

Herein we found three species of florivorous Tephritinae, reared from their host plants: *Trupanea jonesi*, *Dictyotrypeta* sp. and *Tetreuaresta* sp. (**Table 1**), not yet reported to Brazil. Two new species (*Dictyotrypeta* sp. and *Tetreuaresta* sp.) were obtained, which will be later described. Additionally, three species of *Melanagromyza* (Agromyzinae: Agromyzidae) were reared from the sampled host astereces (**Table 6**), probably, are also new species.

The most frequent species of Tephritinae were *Trupanea jonesi* and *Dioxyna chilensis* associated with the Asteraceae: *Porophyllum ruderale*, totaling 41.34% of all florivorous fly species. The least frequent Tephritinae were: *Xanthaciura unipuncta* in *Baccharis triplinervis*, and *Xanthaciura biocellata* in *Eupatorium multicrenulatum*. *Dictyotrypeta* sp. and *Tetreuaresta* sp. were species-specific to flower heads of *Vernonia polyanthes* and *Pterocaulon virgatum*, respectively. The other species of Tephritinae were more generalists, being *Xanthaciura biocellata* the most polyphagous of all the Tephritinae.

Asteraceae Species	**Environments**			**Total %**
	Cerrado	**Semideciduous Forest**	**Agroecosystem**	
Aspilia latissima		*Melanagromyza* sp.3 (1.04%)		1.04
Baccharis triplinervis			*Melanagromyza* sp.2 (1.04%)	1.04
Bidens pilosa	*Melanagromyza* sp.1 (2.08%)	*Melanagromyza* sp.1 (51.04%)	*Melanagromyza* sp.1 (20.83%)	73.95
Bidens sulphurea	*Melanagromyza* sp.2 (3.13%)			3.13
Chaptalia integerrima		*Melanagromyza* sp.2 (5.21%)		5.21
Chromolaena arnottiana	*Melanagromyza* sp.3 (1.04%)			1.04
Emilia fosbergii			*Melanagromyza* sp.2 (1.04%)	1.04
Lourteigia ballotifolia	*Melanagromyza* sp.3 (4.17%)			4.17
Porophyllum ruderale			*Melanagromyza* sp.1 (1.04%)	1.04
Sonchus oleraceus		*Melanagromyza* sp.3 (1.04%)		1.04
Vernonia cognata	*Melanagromyza* sp.3 (1.04%)			1.04
Vernonia polyanthes	*Melanagromyza* sp. 3 (1.04%)	*Melanagromyza* sp.3 (5.21%)		6.25

Table 7.
Frequency of occurrence of Melanagromyza spp. (Agromyzinae, Agromyzidae) associated with Asteraceae species (Asterales) in three phytophysiognomies in Dourados region-MS, mid-west of Brazil (January 2011 to august 2012).

The Asteraceae were more abundant in the Cerrado biome. This same pattern was also accompanied by the species of florivorous fly (N = 374). In the Semideciduous forest there was a highest abundance of *Melanagromyza* species (Agromyzinae), with 802 adults (of three morphospecies), and also of parasitoids (Hymenoptera), being recovered 1,073 individuals.

The Asteraceae *Pterocaulon virgatum* and the tephritine *Tetreuaresta* sp. presented a highly specie-specific relationship. The Asteraceae species with highest abundance of Tephritidae, were: *Baccharis triplinervis*, *Zinnia elegans*, *Eupatorium multicrenulatum*, and *Vernonia polyanthes* that were associated with the tephritines: *Xanthaciura unipuncta*, *Xanthaciura biocellata*, *Dictyotrypeta* sp., and *Cecidochares fluminensis*. The flower heads of *Chaptalia integerrima*, *Praxelis pauciflora*, *Porophyllum ruderale* and *Bidens pilosa* were infested by *Dioxyna chilensis*, and *Trupanea jonesi*. From the host plant *Pterocaulon virgatum* only *Tetreuaresta* sp. emerged.

The most frequent and abundant Tephritinae species in the Cerrado, were: *Xanthaciura unipuncta*, *Tomoplagia brasiliensis*, *Dictyotrypeta* sp., and *Tomoplagia reimoseri*. In the Semideciduous Forest, occurred: *C. fluminensis*, *D. chilensis*, and *Tetreuaresta* sp. The Agroecosystem had the low diversity, occurring only three florivorous flies: *T. jonesi*, *Melanagromyza* sp.1 and *Melanagromyza* sp.2 (**Table 6**).

Some Asteraceae species, such as *Bidens pilosa*, *Porophyllum ruderale*, *Conyza bonariensis*, are invasive plants ("weeds"), that compete with plants grown in agroecosystems. Thus, this study recorded that species of Tephritinae (Tephritidae) and *Melanagromyza* spp. (Agromyzinae, Agromyzidae) feed on the seeds of these invasive plants in their larval phase, having the potential to act in the biological control of these Asteraceae in agrossilvipastoral (agriculture and pasture) areas.

Bidens pilosa, was the species of Asteraceae with greater abundance and frequency of *Melanagromyza* spp., representing 73.95% of the Agromyzidae in this Asteraceae. In the others Asteraceae species, the frequency of occurrence of *Melanagromyza* species was equal to or less than 7% (**Table 6**).

Herein, three species of florivorous tephritines: *Trupanea jonesi*, *Dictyotrypeta* sp. and *Tetreuaresta* sp., are for the first time reported in Brazil. *Dictyotrypeta* sp. and *Tetreuaresta* sp. are two new species that will be later described. Three *Melanagromyza* species (Agromyzinae, Agromyzidae) were recovered from the sampled hosts (**Table 6**), which are also new records and, probably, new species.

The insects that live in plant flowers represent a very sophisticated interaction, because in addition to obtaining physical protection, they obtain a higher quality food (proteins and carbohydrates). This is the first study of trophic interactions between the tephritines, agromizynes, asteraces and parasitoids in flower heads of asteraceae in the Midwest of Brazil.

7. Conclusions and perspectives

1. In the Midwest Brazil occur, at least, 12 species of florivore fly species (9 Tephritinae, and 3 *Melanagromyza*, Agromizinae, Agromyzidae);
2. All Tephritinae and Agromizinae were reared of their Asteraceae host plant flower heads, from three different Biomes (Atlantic Forest, Cerrado, and Agroecosystem);
3. Cerrado is the biome with higher species richness (S = 11) of florivorous fly species, but in the Atlantic Forest occurred higher abundance of their parasitoids (Hymenoptera).
4. Further researches are in need for a better understanding on the resouce partioning between Tephritinae (Tephritidae), *Melanagromyza* spp. (Agromyzinae, Agromyzidae), their association with Asteraceae and their respective hymenopteran parasitoids.
5. Some florivore fly species, such as *Trupanea jonesi*, *Tomoplagia brasiliensis*, *T. reimoseri*, *Xanthaciura biocelata*, *X. unipuncta*, *Dioxyna chilensis* (Tephritinae, Tephritidae), and *Melanagromyza* sp.1 (Agromyzinae, Agromyzidae), need more research on their biology and behavior to be employed in biological control programs against invasive Asteraceae species.

Acknowledgements

We thank the *Conselho Nacional de Desenvolvimento Científico e Tecnológio* (CNPq) for a research grant (Process No. 305112/2012–0) based on research productivity awarded to M.A. Uchoa, and a MSc grant to Morgana Fracieli Wachter-Serapião, and Universidade Federal da Grande Dourados (UFGD), for the logistic and financial support.

Conflict of interest

The authors declare no conflict of interest.

Author details

Manoel A. Uchoa[1*], Morgana F. Wachter-Serapião[1] and Nádia Roque[2]

1 Laboratory of Taxonomy and Systematics of Tephritidae (LabTaxon), Universidade Federal da Grande Dourados (UFGD), Dourados, MS, Brazil

2 Universidade Federal da Bahia (UFBA), Salvador, BA, Brazil

*Address all correspondence to: uchoa.manoel@gmail.com

References

[1] Norrbom, A L (2010). Tephritidae (fruit flies, moscas de frutas), p. 909-954. *In*: Brown BV, Borkent A, Cumming JM, Wood DM, Woodley NE, Zumbado, MA (eds.). **Manual of Central American Diptera**. NRC Research Press, Otta. Vols. 1+ 2: 1.442p.

[2] Prado PI, Lewinsohn TM, Almeida AM, Norrbom AL, Buys BD, Macedo AC, Lopes MB (2002). The fauna of Tephritidae (Diptera) from capitula of Asteraceae in Brazil. Proc Entomol Soc Wash 104:1007-1028.

[3] Uchoa, MA (2012). Fruit Flies (Diptera: Tephritoidea): Biology, host plants, natural enemies, and the implications to their natural control, pp. 271-300. *In*: Larramendy, M. L. & Soloneski, S. (eds.). **Integrated Pest Management and Pest Control - Current and Future Tactics.** ISBN 978-953-307-926-4. DOI 10.5772/31613. InTechOpen. Rijeka, Croatia. 668p. https://www.intechopen.com/books/integrated-pest-management-and-pest control-current-and-future-tactics/fruit-flies-diptera-tephritoidea-biology-host-plants-natural-enemies-and-the-implications-to-their-n (Accessed on January 05, 2021).

[4] Funk VA, Susanna A, Stuessy TF, Bayer RJ (eds.) (2009). **Systematics, evolution, and biogeography of compositae**. IAPT-International Association for Plant Taxonomy, Vienna, Austria. 965p.

[5] Bellard C, Leclerc C, Leroy B, Bakkenes M, Veloz S, Thuiller W, Courchamp F (2014). Vulnerability of biodiversity hotspots to global change. Global Ecol. Biogeogr. 23 (12): 1376-1386. DOI: 10.1111/geb.12228.

[6] https://agenciadenoticias.ibge.gov.br/agencia-sala-de-imprensa/2013-agencia-de-noticias/releases/29367-em-2014-mata-atlantica-era-o-bioma-com-maior-numero-de-especies-ameacadas-de-extincao#:~:text=A% 20pesquisa%20%E2%80%9 CContas%20de%20Ecossistemas, 1.989%20estavam%20na% 20Mata%20Atl%C3%A2ntica (Accessed on Jan 17, 2021).

[7] Wachter-Serapião MF, Uchoa MA (2013). Diversidade de Moscas Florívoras (Diptera: Tephritoidea) e suas interações com espécies de Asteraceae e Parasitóides (Hymenoptera) em três paisagens antropisadas no Brasil. Dissertação de Mestrado. Programa de Pós-Graduação em Entomologia e Conservação da Biodiversidade, Universidade Federal da Grande Dourados (UFGD), Dourados-MS, Brazil. 32p.

Chapter 6

Chironomidae: Biology, Ecology and Systematics

Zerguine Karima

Abstract

The family of Chironomidae is a group of Diptera insects belonging to the suborder of Nematocera, commonly called "non-biting midges" in the adult stage and "bloodworms" in the larval stage. The Chironomidae are often the most abundant group of macroinvertebrates, in number of species and individuals, encountered in all aquatic environments of freshwater, brackish, terrestrial and even the sea. Likewise, Chironomidae occur in all the continents. The Chironomidae family is divided into 11 sub-families that have diffrent ecological statues. Despite the wealth of data on Chironomidae in the Holarctic region, other parts of the world are poorly studied and few guides to identifying Chironomidae have been produced. This chapter includes a theoretical synthesis on the Chironomidae, it deals with the Biology (life cycle and description of different stages), description of all subfamilies and the ecology of this important family of Diptera.

Keywords: Biology, Chironomidae, Diptera, ecology, subfamilies

1. Introduction

The Chironomidae family is a group of Diptera insects belonging to the suborder of Nematocera. Members of this family are commonly called "non-biting midges" in the adult stage and "bloodworms" in the larval stage.

The Chironomidae are often the most abundant group of macroinvertebrates, in number of species and individuals, found in all freshwater aquatic environments. They are widely distributed and live in both lentic and lotic ecosystems [1, 2]. Indeed, the Chironomidae are among the few insects living in the sea and the ocean [3, 4]. Likewise, they occur in all continents where they have been found alive at heights of 5600 m on the glaciers of the Himalayas [5] and in the depths of the lakes [6]. Several qualitative observations showed that larvae of terrestrial Chironomidae are able to colonize the vegetation above the soil surface on heathlands [7].

Chironomidae are holometabolous insects, their larvae, pupae and adults form an integral part of the trophic chain serving as food for other invertebrates, fish, birds and amphibians [8, 9]. The larval and pupal stages are generally subservient to aquatic habitats while the adults are aerial and often collected at more or less distances from their emergence habitats.

The Chironomidae family is divided into 11 sub-families: Telmatogetoniinae, Usambaromyiinae, Podonominae, Tanypodinae, Buchonomyiinae, Diamesinae, Prodiamesinae, Orthocladiinae, Chironominae, Chilomyiinae and Aphroteniinae. In fact, the subfamilies of Telmatogetoniinae, Podonominae, Buchonomyiinae,

Chilomyiinae, Usambaromyiinae and Aphroteniinae are restricted in their distributions and even in number of species. The occurrence of Diamesinae and Prodiamesinae depends on climatic conditions. However, the subfamilies of Orthocladiinae, Tanypodinae and Chironominae are those which encompasses the maximum number of species and are very widely distributed throughout the world [2, 10].

Among the Chironomidae family, description of species is traditionally based on adults, and knowledge of immature stages is variable within tribes or even across species of the same genus. Indeed, some genera have immature terrestrial stages, other genera have exclusively aquatic larvae. However, many species have unknown immature stages [11].

Regional catalogs provide valid data on the distribution of Chironomidae: [12] for the Eastern region, [13] for the Afrotropical region, [14] for the Australian and Antarctic regions, [15] for Nearctic Chironomidae, [16] for the Neotropical region, [17] for Europe and [18].

The common problem in the literature dealing with the morphology of Chironomidae is that several alternative names are frequently used for the same structure. In fact, [19] glossary is a constructive attempt to rationalize this situation. For this reason, his recommendations have been followed throughout this chapter.

The objective of this work is to present an overview on the Chironomidae family. This chapter will mainly deal with the morphology of all the subfamilies as well as the biology and ecology of the different stages.

2. Morphology

Chironomidae are Diptera belonging to the morphological group of the Culiciforma, so their general appearance is that of a mosquito. They are Nematoceran and as such, they are characterized by long antennae (more or less as long as the head). Their mouthparts are much regressed and the atrophy of the mandibles in the adult stage does not allow them to bite.

Chironomidae undergo during their life cycle four morphologically very different stages which, while having a general appearance identical from one subfamily to another, present anatomical variations which constitute essential bases of their systematics.

2.1 Eggs

2.1.1 Structure

The egg of Chironomidae, like all insects, is of the centrolecithic type, rich in yolk which constitutes a central mass of nutrient reserves. The cytoplasm containing several nuclei is peripheral [20].

The eggshell has, from the inside to the outside, the yolk envelope and the chorion separated by a protective waxy layer. In general, the chorion of eggs of Chironomidae is not very thick and contains protrusions and has a micropyle [21]. However, it may be smooth in other species such as *Tanytarsus barbitarsis* [22] or thick providing some protection against desiccation in eggs of Telmatogetoninae [23].

In general, all Chironomidae laid their eggs in the form of gelatinous masses in contact with water. However, members of the Telmatogetoninae subfamily are an exception since their eggs are laid individually without a gelatinous matrix [23].

2.1.2 Number of eggs

Often, the egg masses of Chironomidae contain approximately 20 to 30 eggs. This number can increase to over 3000 in large species [24]. In fact, the largest number of eggs laid was recorded in *Chironomus tentans* with 3300 eggs in a single mass. However, there may also be intraspecific variations [23].

2.1.3 The shape and size of the eggs

The shape of the eggs in Chironomidae is usually elliptical or kidney-shaped. Likewise, the eggs can also be deltoids in some Telmatogetoninae (*Telmatogelton japonicus*) and some Orthocladiinae such as *Orthocladius* sp. and *Eukiefferiella claripennis* [23].

Egg sizes vary greatly between species. Indeed, the smallest eggs are those of *Corynoneura* and *Thienemanniella* whose size is around 170 μm long and 70 μm wide, while *Tanypus punctipennis*, a large Tanypodinae, lays eggs 612 μm long and 135 μm wide. Generally, in Chironomidae the ratio: length/width is 2.5 to 3 [23].

Egg masses may be globular or rod-shaped in Tanypodinae, laid as chains in Diamesinae, or linear in Orthocladiinae. In Chironominae, Chironomini egg masses are cylindrical in shape with a gelatinous peduncle (**Figure 1**) [24, 25].

2.1.4 Embryonic development

The duration of embryonic development is largely influenced by environmental factors especially temperature [26]. In fact, *Thiennemanniella vittata* eggs hatch in a minimum of 4 days at 20°C, 6 days at 15°C, 13 days at 10°C and 31 days at 5°C [21].

2.2 The larvae

The Chironomidae undergo four larval stages but all morphological and taxonomic observations have been made on the last stage. The majority of structures appear in the early larval stages [27] but many characters of the final stage, especially the shapes and ratios, do not apply to the early stages and do not allow good differentiation [28, 29].

The larvae of Chironomidae have a well-individualized, developed, exposed, complete, and non-retractile head capsule and a narrow, elongated segmented body that lacks thoracic legs (**Figure 2**) [28].

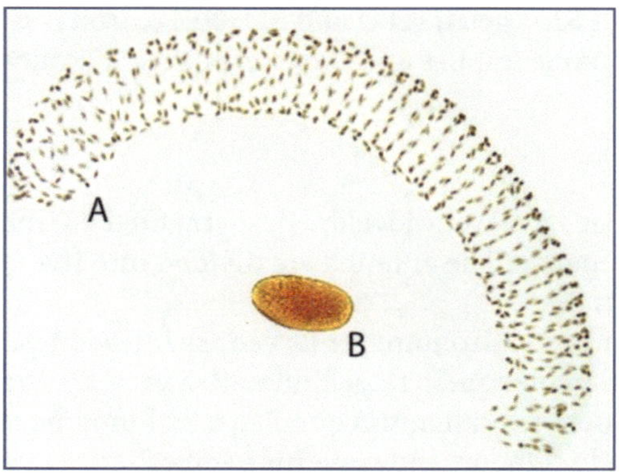

Figure 1.
Chironomus striatipennis. *(A): Egg mass; (B): Egg [25].*

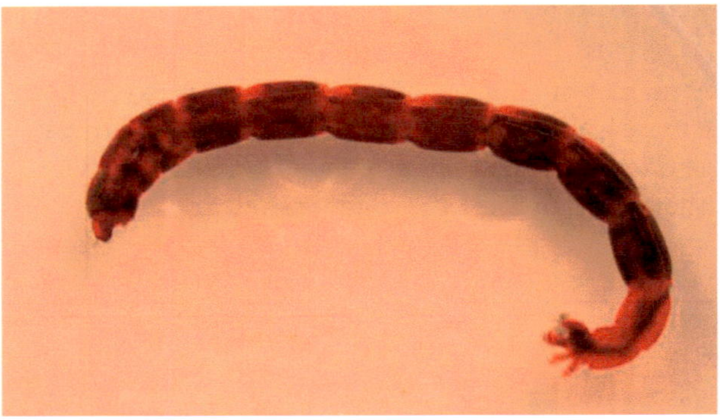

Figure 2.
Chironomidae larva [30].

The larvae of Chironomidae are almost 3 to 25–30 mm long. Their coloration is variable ranging from whitish yellow to red, green or blue. They can also be brownish, purple or orange. Sometimes there is ornamentation on the body segments.

The main parts of the larva's body are:

2.2.1 The cephalic capsule

For the identification of Chironomidae larvae it is necessary to know the morphological details, especially on the head and the perianal region.

2.2.1.1 The cephalic skeleton

The head capsule of Chironomidae consists of a sclerotinized cranium, which consists of the dorsal apotome and a pair of lateral genae. These three sclerites are separated by sutures. There are morphological differences between the different subfamilies but in the most divided state, the clypeus is inconspicuous and more than five labral sclerites are found anterioriorly to the frontal apotome. Variations can occur and include the fusion of the clypeus and the frontal apotome to form the frontoclypeal apotome (**Figure 3**).

The position of the dorsal cephalic seta is related to these sclerites, in fact, S1 and S2 are found on the labrum (labral seta), S3 on the clypeus (clypeal seta) and S4 and S5 on the frontal apotome (frontal seta).

The genae form the lateral and ventral parts of the sclerotinized head and on which are the remaining seta: S6 (suborbital seta), S7 (supraorbital seta), S8 (parietal seta), S9 and S10 (genal seta) and S11, S12 (coronal seta). All of these seta are localized as their name implies and they are of great taxonomic value [1].

2.2.1.2 Antenna

Most Chironomidae have well-developed, segmented antennae placed anterodorsally on the upper genae. The antenna are divided into five segments but there can be 4 or 3 or 7 segments.

Usually the antenna of Chironomidae larvae are formed by a basal segment of varying length and diameter and a flagellum with a varying number of segments. On the the basal segment we can note a circular sensillum: the annular organ. On the apex of the second segment, and sometimes on others, is a special sensory-function formation: the Lauterborn organs, which may be sessile or pedunculated.

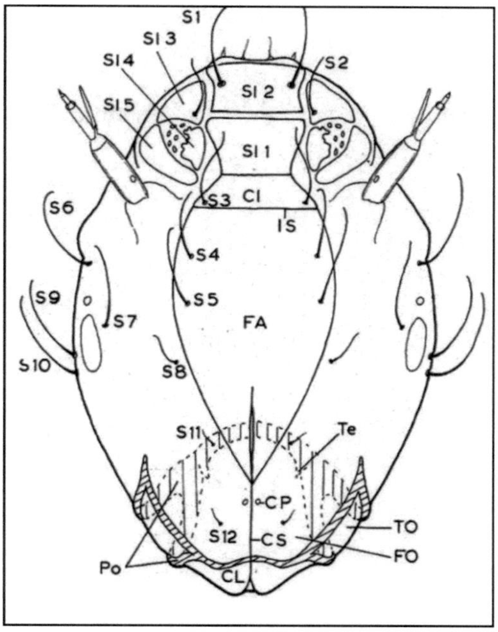

Figure 3.
Cephalic capsule of a Chironominae (dorsal view). Cl: clypeus, CL: coronal lobe, CP: coronal pores, CS: coronal suture, FA: frontal apotome, FO: occipital foramen, IS: clypeolabral suture, Po: postmentum, S 1–12: cephalic seta, Sl (Sl 1–5): sclerites. Te: tentorium, TO: occipital triangulum [19].

2.2.1.3 Labrum

The labrum represents the anterior portion of the frontal apotome. The ventral surface of the labrum is the epipharynx or palatum, which bears seta, lamellae or spinula, it has a sensory and nutritional role (**Figure 4**).

On the labrum are inserted the posterior labral seta named: SIVA and SIVB. SIVA are large sensilla and the SIVB are smaller and may be missing. More anteriorly, there are three pairs of seta: the SIII which are fine simple and in a mid-posterior position compared to the SII seta. The latter are also often simple but can be large and pectinate. The most posterior labral setae are the SI which are very variable and have an important taxonomic role (especially in Orthocladiinae). Indeed, they can be simple or bifid, fluffy or pectinate.

In the ventral surface of the labrum are inserted the premandibles which are mobile appendages bearing one or more apical teeth. They are with or without a tuft of seta called: the premandibular brush.

2.2.1.4 Mandibles

The mandibles are mouthparts. They are toothed with a dorsal external tooth which is missing in the majority of taxa, an apical tooth and a variable number of internal teeth (often 2–3) (**Figure 5**).

Three seta or groups of seta can be identified:

- the pecten mandibularis: which has the shape of a comb, it is located on the subapical surface.
- the seta subdentalis on the internal mola.

- the seta interna, basal, branch-like, often located on the internal mandibular surface.

The mandibles can be sickle-shaped in Tanypodinae or larger in other subfamilies.

2.2.1.5 Mentum or Labium or inferior labrum

The mentum is a mid-central sclerotized part of the cephalic capsule almost always provided with teeth. This piece has two walls: the dorsomentum and the ventromentum, the latter may extend laterally in ventromental or paralabial plates (**Figure 6**).

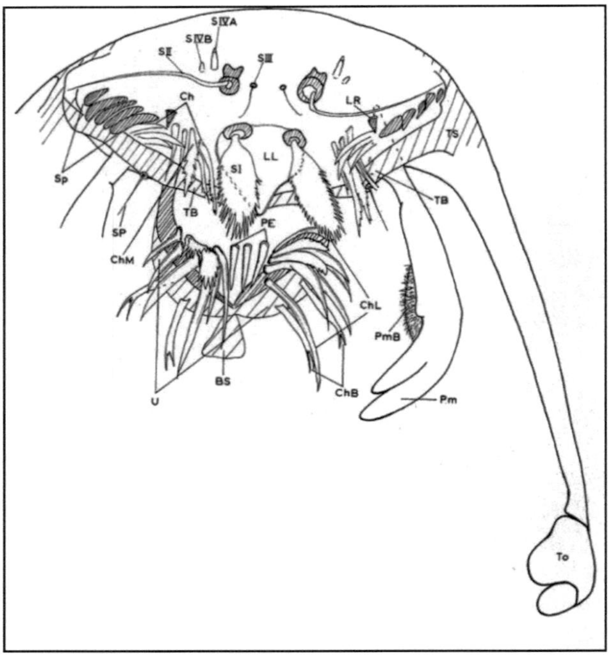

Figure 4.
Structure of the labrum. BS: basal sclerite, Ch: chaeta, ChB: basal chaetulae, ChL: Chaetulae laterales, ChM: media chaetules, LL: lamella labrales, LR: labral rod, PE: pecten epipharyngis, Pm: premandibles; PmB: premandibular brush; SI, SII, SIII, SIVA, SIVB: labral setae, SP: seta premandibularis, TB: tormal bar, To: occipital triangulum, TS: triangular sclerite, U: ungula [19].

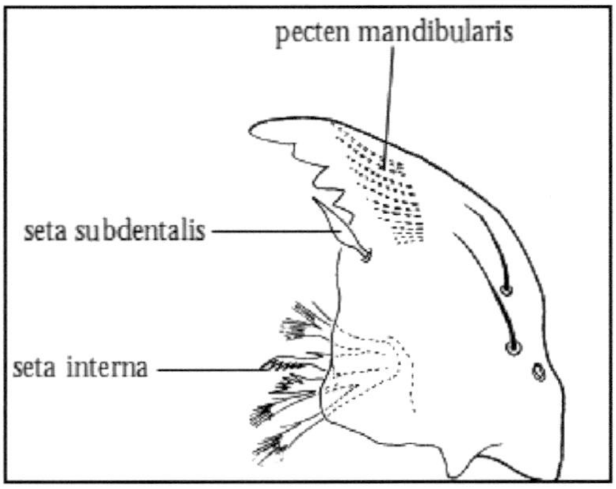

Figure 5.
Structure of the mandible of Chironomidae [29].

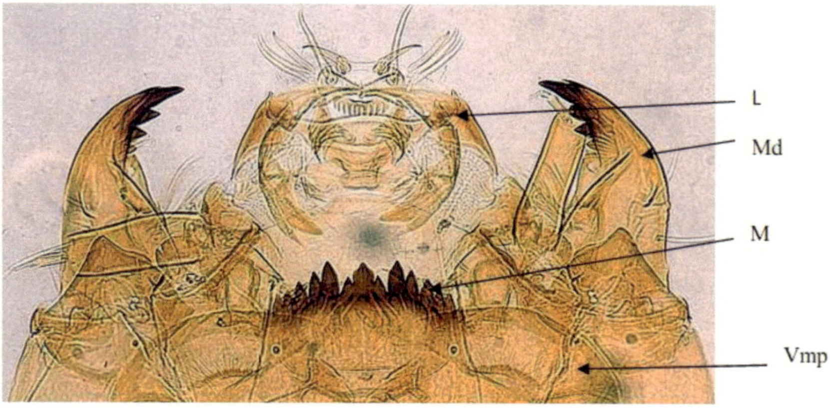

Figure 6.
Structure of the mentum of Chironominae. L: labrum, M: mentum, Md: mandible, Vmp: paralabial plates [31].

The presence or absence of these paralabial plates, in addition to their shape, is of great systematic importance. Indeed, in Tanypodinae the ventromentum is hyalin and the dorsomentum appears as a toothed surface. In the Orthocladiinae subfamily, the ventromental plates are reduced. In Chironominae, they are highly developed and their dorsal surface is variably striated.

2.2.1.6 Premento-hypopharyngial complex

The Premento-hypopharyngial complex lies dorsally to the mentum, and is completely covered by it in the ventral position. It consists of two lobes:

- The prementum: ventral, it is well developed in Tanypodinae. In the other subfamilies, these structures are very small and differently constructed.
- The hypopharynx: dorsal and never well developed. It has several rows of scales or, in Tanypodinae, has rows of teeth on each side forming the hypopharyngial pecten.

2.2.1.7 Eyes spots

The eyes of the larvae of Chironomidae are simple subcuticular areas of pigment. There are taxonomic differences regarding the shape and position of eyes spots. In fact, in the larvae of the Chironominae there are two vertical eyes spots, while those of the Orthocladiinae are also double and superimposed on each other (**Figure 7**). In Tanypodinae there is a single eye spot on each side and are kidney shaped. In all the other subfamilies the eyes spots are simple.

2.2.2 The body

The body is divided into a thorax and an abdomen (**Figure 8**).

- Thorax: always with three segments. The only appendages found in the thorax are two non-segmented pseudopods with a crown of simple or hook-shaped claws.
- Abdomen: narrower, made up of nine segments. It bears on the terminal segment the posterior pseudopods. On the dorsal side of the penultimate segment there are a pair of tubercles, each supporting a tuft of setae, called: the procercus. In the terminal region of the abdomen there are also often 02 pairs of anal tubules.

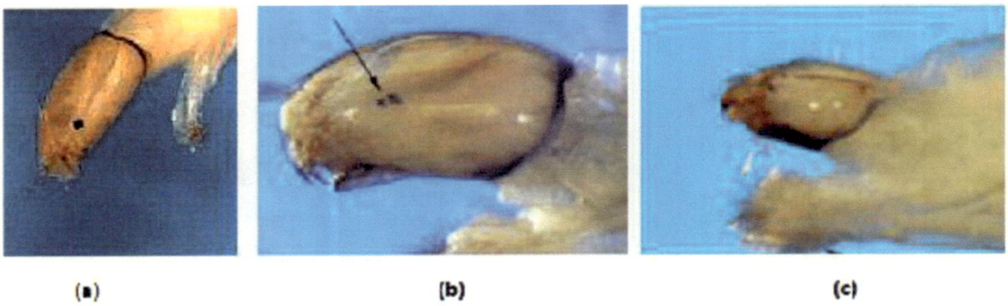

Figure 7.
Position of eye spots in Chironomidae. (a): Tanypodinae, (b): Orthocladiinae and (c): Chironominae [32].

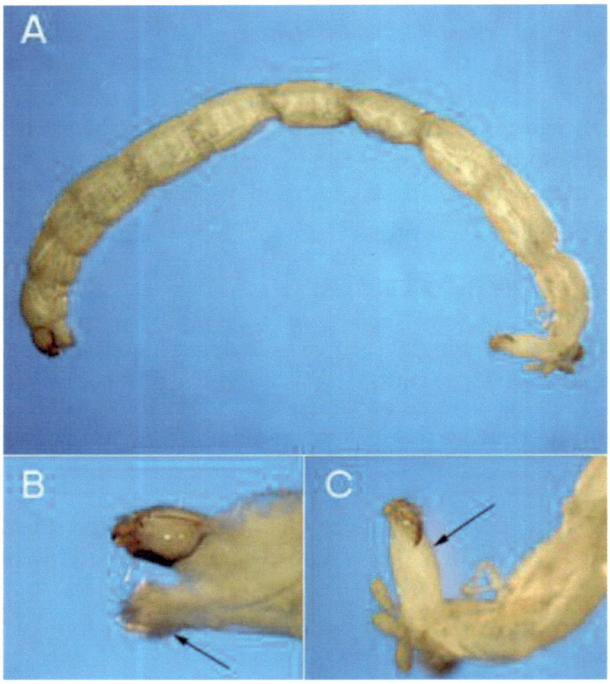

Figure 8.
The body of the larva of Chironomidae. (A): complete larva; (B): anterior pseudopods; (C): anal region of a Chironominae. Arrows show the anterior and posterior pseudopods [32].

2.3 The pupa

The pupal stage of Chironomidae is very short compared to the larval stage, its duration is from a few hours to several days.

The characters of the pupae of Chironomidae are best seen on their exuviae, which are very useful tools for species determination [33, 34].

The pupae of Chironomidae are comma shaped with a swollen cephalothorax and a dorsoventrally flattened abdomen (**Figure 9**). Their length varies from just under 3 to 18 mm. Their coloration usually follows that of the larva.

There are three regions in the pupa: the head (the cephalic region), thorax and abdomen. However, the head and thorax are fused together forming the cephalothorax.

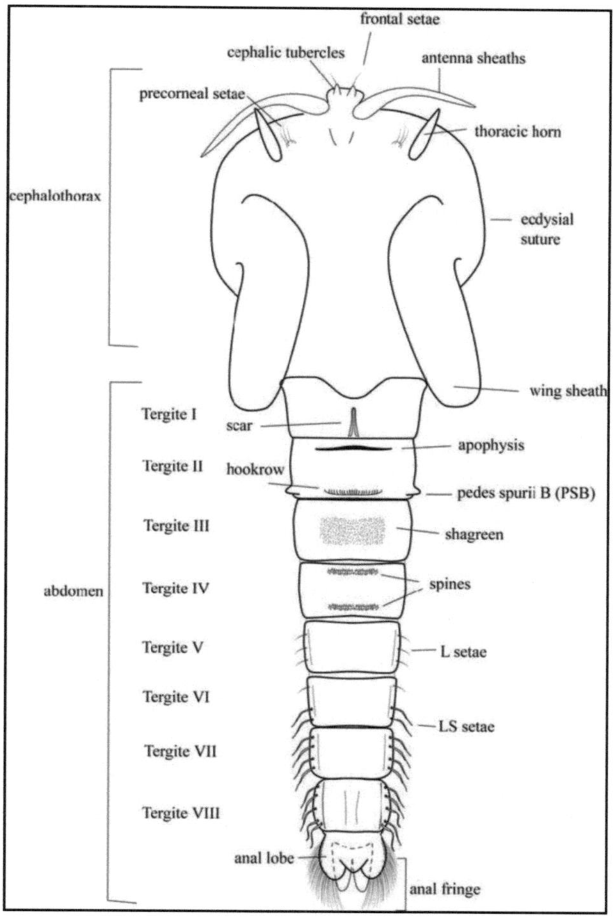

Figure 9.
Morphology of Chironomidae pupa (Illustration by M.R. Rufer).

2.3.1 Cephalothorax

The head bears the antennal sheaths, a pair of cephalic tubercles and a pair of frontal setae. Of great taxonomic importance are features of the cephalic region: location, number and length of frontal setae, the vertex, the postorbit and the ocular region.

The thorax has a pair of respiratory organs, also called: prothoracic horns or thoracic horns, which vary greatly in shape depending on the species or genera, these organs serve for respiration. The thorax also carries the wing sheaths, or pterotheca and the legs sheaths or podotheca.

2.3.2 Abdomen

The abdomen of Chironomidae pupa is made up of nine articulated segments. In addition to a number of setae, the pupal exuvia exhibits highly variable ornamentation (spines, spinules and tubercles) useful for the characterization of the various taxonomic levels.

The last segment widens forming the two anal lobes. The external margin of these anal lobes always bears swimming setae forming the swimming fringe. In addition to the swimming setae, near the apex of the last segment there are often thick and curved setae, in number of 3 or 4, which are called the apical and subapical setae. These can also be missing. In their distal part, the anal lobes bear filamentous setae called: anal macrosetae.

2.4 The adult (the imago)

The body of the adult of Chironomidae consists of three parts [35]:

2.4.1 The head

The head: globular, it carries:

- The antennae: long and exhibit sexual dimorphism since they are fluffy in males and moniliform in females. The antenna of the adult Chironomidae consists of a narrow scape, a globose pedicel and a number (often 11–14) of flagellomeres. The number of antenna segments and their shape depend on the species [36].

- The eyes are very large and kidney-shaped.

- The mouthparts are very reduced.

2.4.2 Thorax

Thorax generally well developed, it has three parts of equal importance: pronotum, mesonotum and metanotum. The thorax bears the wings and the legs.

2.4.3 Abdomen

Abdomen composed of 10 segments, the seven anterior segments are flattened dorsoventrally. The female's abdomen is shorter and more swollen than that of the male. The dorsal part has coloring or ornamentation often useful for identification. The last abdominal segments form the genitalias. The tergite IX has a posteromedial extension forming the anal point (**Figure 10**). Among the most distinctive characters of male genitalias are: basal gonocoxites and apical or subapical gonostyles. The gonocoxites of

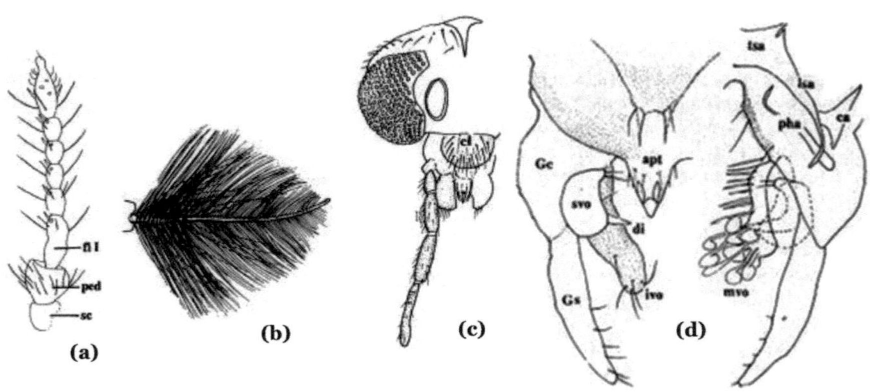

Figure 10.
Morphology of the adult: (a): antenna of the female; (b) the male antenna; (c): the head; (d): male genitalia. Apt: anal point; ca: coxapodema; cl: clypeus; di: digitus; fl: flagellomere; Gc: gonocoxitis; Gs: gonostyle; ivo: inferior volsella; lsa: lateral sternapodema; mvo: median volsella; ped: pedicel; pha: phallapodema; sc: scape; svo: superior volsella; tsa: transverse sternapodema [1, 35].

Chironomidae support a varying number of appendages called: volsellae and they are named according to their relative positions (middle, inferior and superior). Likewise, there may be other lobes associated with the aedeagus and the penis.

3. Subfamilies of the Chironomidae family

The Chironomidae family is divided into 11 subfamilies and 22 tribes revised and provided by [37]. The subfamilies of the Chironomidae are: Telmatogetoninae, Podonominae, Tanypodinae, Buchonomyiinae, Diamesinae, Prodiamesinae, Orthocladiinae, Chironominae, Chilenomyiinae, Aphroteniinae and Usambaromyiinae (**Table 1**).

3.1 Subfamily of Orthocladiinae

The Orthocladiinae subfamily includes species characterized at the adult stage by the fact that the distal article of the hypopygium (gonostyle) is folded over the proximal article (gonocoxite). Likewise, the first section of the tarsus of the first pair of legs is shorter than the tibia, therefore the tibio-tarsal ratio (or leg ratio) is less than 1 (**Figure 11**).

Sub families	Tribes
Tanypodinae	•Coelotanypodini •Macropelopiini •Pentaneurini •Tanypodini •Anatopyiini •Coelopyiini •Natarsiini •Procladiini
Podonominae	•Boreochlini •Podonomini
Telmatogetoninae	
Diamesinae	•Diamesini •Protanypodini •Boreoheptagyiini •Harrisonini •Heptagyiini •Lobodiamesini
Prodiamesinae	
Orthocladiinae	•Metriocnemini •Orthocladiini •Corynoneurini
Chironominae	•Chironomini •Tanytarsini •Pseudochironomini
Buchonomyiinae	
Aphroteniinae	
Chilenomyiinae	
Usambaromyiinae	

Table 1.
Major divisions of the Chironomidae family [37].

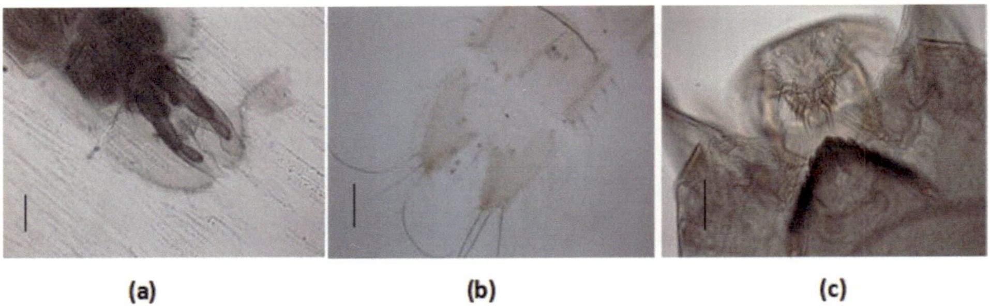

Figure 11.
Orthocladiinae subfamily (a): anal lobe and hypopygium of Psectrocladius *(*Allopsectrocladius*)* platypus; *(b): anal lobe of* P. *(*P.*)* limbatellus; *(c): cephalic capsule of the larva of* Cricotopus flavocinctus *[38].*

Thoracic horn can be absent in the pupae. When it is present it never be branched. On the posterolateral angle of the tergites there are never any hooks. In some genera the anal lobe have a swimming fringe.

The larvae (3 to 15 mm) are always devoided of ventral tubules and hemoglobin, therefore their coloring is never reddish. Usually they are white or yellowish, sometimes greenish, brown or purple. The ventromental plates are in general narrow and never developed.

3.2 Subfamily of Tanypodinae

Tanypodinae are Chironomidae characterized by antennae with 15 segments in adult males and 11–15 segments in females. Males are distinguished from those of Diamesinae and Prodiamesinae by the morphology of their hypopygium having a simple structure and lacking basal lobes at the coxites (**Figure 12**).

The pupae have generally extended anal lobes always provided with 02 large lateral setae which have a swimming function. In the majority of species the respiratory organs end distally in a sieve plate.

The larvae are usually slender with long procerci and pseudopods and anal tubules. What characterizes the larvae of Tanypodinae is the presence of relatively long retractile antennae, the eyes spots are kidney-shaped and characteristic mouthparts (the premento-hypopharyngial complex).

3.3 Subfamily of Chironominae

The Chironominae subfamily includes species characterized in the adult stage by a reduced anterior tibial spur, genitalia most often having a complex structure. The gonostylus cannot be folded up and are always in the prolongation of the coxites.

The respiratory horn may be lacking in pupae, if they exist they can be fluffy or simple. The tergites are often ornamented with hooks. Numerous sub-equal setae are also present in swimming fringe (**Figure 13**).

The larvae of Chironominae are generally provided with hemoglobin and colored red. They are easily recognized by the arrangement of their eyes spots, the paralabial plates which are always well developed and striated and the presence, in many species, of ventral tubules in the anal region.

3.4 Subfamily of Diamesinae

The Diamesinae are characterized in the adult stage by the morphology of the male hypopygium and the antennae. They are distinguished from the Prodiamesinae by the bifurcation of the cubital vein. Likewise, tarsomere 4 is cordate.

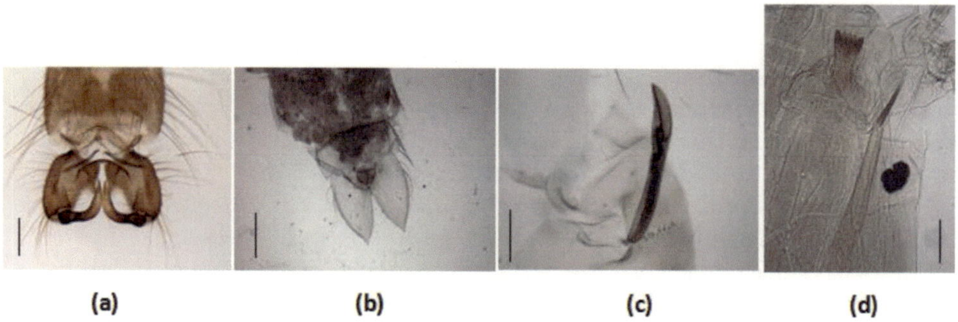

Figure 12.
Tanypodinae subfamily (a): male hypopygium of Arctopelopia melanosoma*; (b): anal lobe of the pupa of* Psectrotanypus varius*; (c):* thoracic *horn of* Xenopelopia falcigera*; (d): cephalic capsule of the larva of* X. falcigera *(scale bar: 200 µm) [38, 39].*

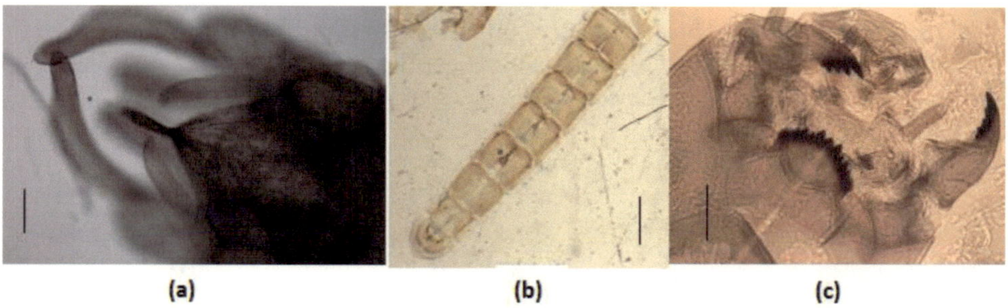

Figure 13.
Subfamily of Chironominae (a): hypopygium of Chironomus plumosus *(scale bar 200 µm); (b): pupal exuvia of* Glyptotendipes barbipes *(scale bar 200 µm); (c): mentum and mandible of* Endochironomus tendens *(scale bar 100 µm) [38].*

Diamesinae pupa often have a thoracic horn which is very variable in shape but it may be lacking in some genera such as *Pseudodiamesa* and *Pottasia* (**Figure 14**).

The larvae are characterized by the presence of a large dark occipital border and the third antenna segment which is annular.

Among the species belonging to the Diamesinae subfamily: *Pseudodiamesa nivoka, Potthastia longimanus, Diamesa steinboecki, Diamesa aberrata, Boreoheptagyia* sp.

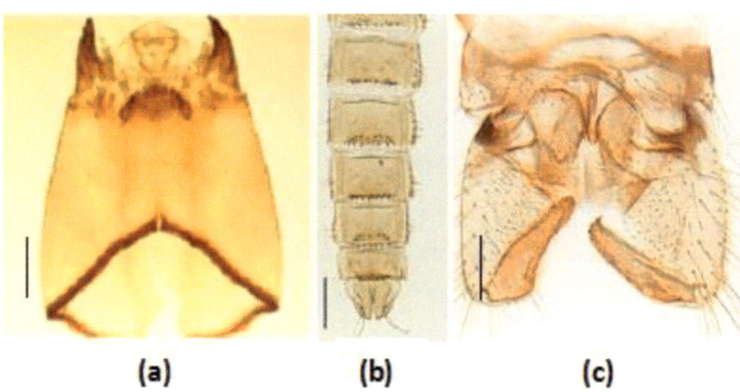

Figure 14.
Diamesinae subfamily: (a): the cephalic capsule; (b): pupal exuvia; (c): male hypopygium [39, 40].

3.5 Subfamily of Prodiamesinae

The Prodiamesinae are characterized by the fact that in the adult stage the wings have, like the Diamesinae and the Tanypodinae, the mediocubital cross-vein, but the difference is that the bifurcation of the cubitus is in a distal position with respect to the radio-median and the medio-cubital cross-veins.

The pupae of Prodiamesinae are very similar to those of Orthocladiinae, but they are distinguished mainly by the thoracic horn which is always present.

Prodiamesinae larvae are characterized by antennae constituted by four segments. The basal segment is much larger than the three successive ones. At the apex of the second segment there are two Lauterborn organs (**Figure 15**).

Among the species belonging to the Prodiamesinae subfamily: *Monodiamesa tuberculata, Prodiamesa olivacea, Odontomesa fulga.*

3.6 Subfamily of Podonominae

Podonominae adults are characterized by wings with the mediocubital cross-vein more distal than in Prodiamesinae. Male genitalia are very variable, in fact, gonostyles are either simple (*Lasiodiamesa sphagnicola*) or bilobed (*Parochlus kiefferi*). Larvae are characterized by their antennae in which the segment 3 is often with annulations (**Figure 16**).

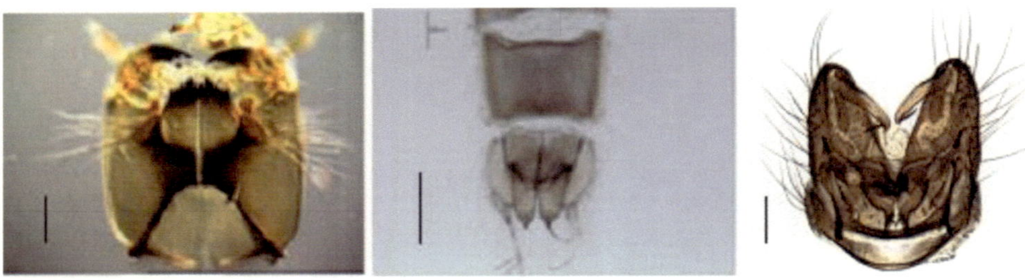

(a) (b) (c)

Figure 15.
The Prodiamesinae subfamily. (a): cephalic capsule; (b): pupal exuvia; (c): male hypopygium (scale bar: 200 μm) [41, 42].

(a) (b) (c) (d)

Figure 16.
The Podonominae subfamily. (a): antenna of Boreochlus *sp.; (b): mentum of* Boreochlus *sp.; (c): procerci of* Podonomus amarali*; (d): pupal exuvia of* Podonomus amarali*; (e): wing; (f): male hypopygium of* Podonomus tehuelche *[43–45].*

3.7 Subfamily of Telmatogetoninae

Species belonging to the Telmatogetoninae subfamily are characterized by having wings with R2 + 3 absent. In addition the veins R1 and R4 + 5 are widely separated. Likewise, the fourth tarsal segment is cordate and the male genitalia are large and variously constructed (**Figure 17**).

Among the species belonging to the Telmatogetoninae subfamily: *Telmatogeton remanei, Telmatogeton japonicus, Thalassomyia frauenfeldi, Psammathiomyia pectinata*.

3.8 Subfamily of Buchonomyiinae

Adults of the Buchonomyiinae subfamily are characterized by wings with the M-Cu vein very close to the base. Male sternite IX with small lateral lobes never exceeding the genitalia. The subfamily of Buchonomyiinae consists of a single genus: *Buchonomyia* (**Figure 18a,b**).

3.9 Subfamily of Aphroteniinae

Adults of the Aphroteniinae subfamily are characterized by wings with R2 + 3 absent, in addition R1 and R4 + 5 are widely separated. Likewise, the fourth tarsal segment is cylindrical (**Figure 18c,d**). Among the species of the Aphroteniinae subfamily: *Aphroteniella filicornis.*

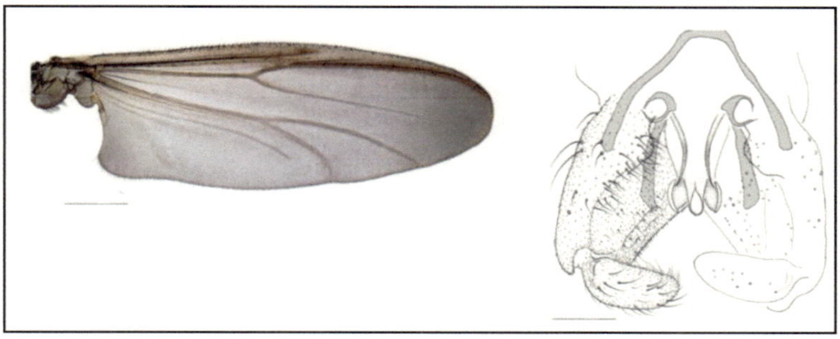

Figure 17.
The Telmintogetoninae subfamily. Wing and male hypopygium of Telmatogeton yamaguchiae *(scale bar: 100 µm) [46].*

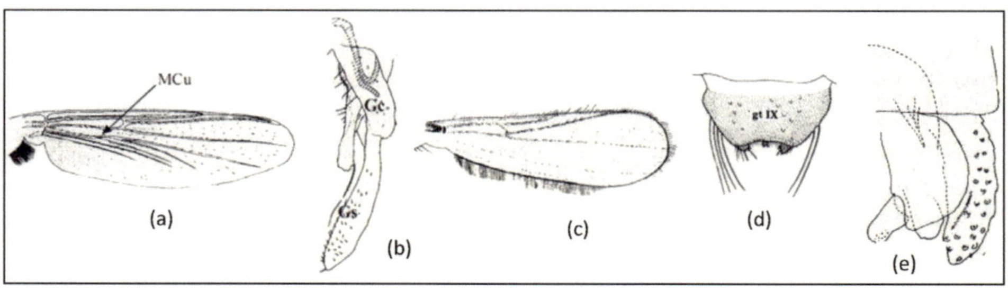

Figure 18.
Subfamilies of Buchonomyiinae, Aphroteniinae and Chilenomyiinae. (a) and (b): wing and hypopygium of Buchonomyiia*; (c) and (d): wing and hypopygium of* Aphroteniella filicornis*; (e): hypopygium of* Chilenomyia *sp. (Gc: gonococcite; Gs: gonostyle; gt: gonotergite) [1].*

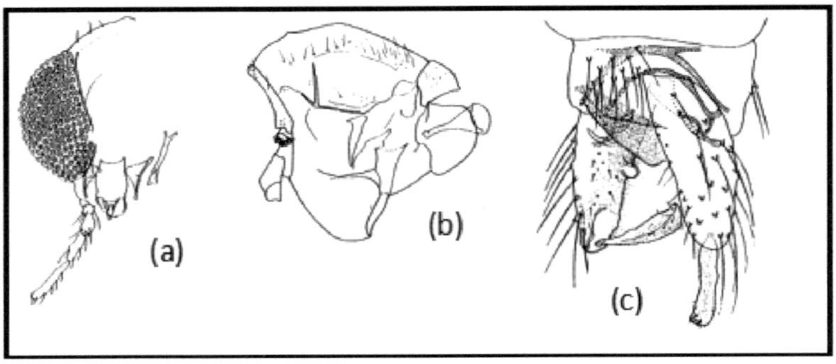

Figure 19.
The subfamily of Usambaromyiinae: (a) the head; (b): thorax; (c): male hypopygium [48].

3.10 Subfamily of Chilenomyiinae

The Chilenomyiinae subfamily is characterized by adults having wings with the mid-cubital vein present, the R1 and R4 + 5 veins are widely separated, in addition, the R2 + 3 is absent. The IX sternites of males are not attached to the IX gonocoxites, but lateral lobes extend further forward than the gonocoxites and gonostyles (**Figure 18e**).

Chilenomyiinae species have been described only from adult males and females. The larvae and pupae of this subfamily are not known.

Among the species belonging to this subfamily: *Chilenomyia paradoxa* [47].

3.11 Subfamily of Usambaromyiinae

This subfamily described by [48] represented by a single species: *Usambaromyia nigrala,* conceived as forming the plesiomorphic sister group of the Tanypodinae, the Podonomiinae and the Aphroteniinae combined.

The species differs from other chironomids by having nearly completely black wings in both sexes, tibial spurs with lateral denticles making them Tanypodinae—like, male gonostylus without megaseta. Pupae and larvae are unknown [48] (**Figure 19**).

4. Biology of Chironomidae

4.1 Life cycle

Chironomidae are Holometabolous insects, their development cycle comprises four morphologically very different states which, while having a general appearance identical from one subfamily to another, present anatomical variations which constitute one of the essential bases of systematics (**Figure 20**).

The life cycle of Chironomidae begins with the deposition of eggs in water. These are gathered in gelatinous masses or deposited individually. The eggs may be free or attached to an object. They hatch after a more or less long period releasing the larvae. The latter undergo a reduced number of moults and go through four larval stages. The larvae can be free, sedentary or live inside a tube that they build with the substrate and salivary secretions.

The larval stage is followed by that of the pupa. This can swim freely or, in tubicolous species, can remain partially included in the larval tube. At maturity, the pupa reach the surface of the water with the air produced in the intercuticular space of

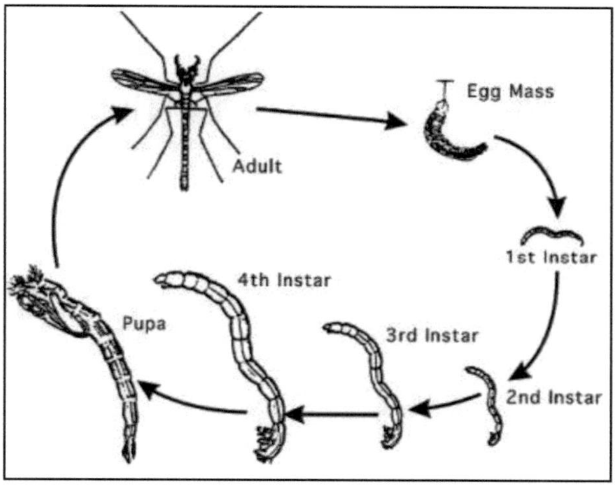

Figure 20.
Chironomid life cycle [49].

the adult. The adult emerges from the surface of the water in a very short period of time. Imaginal life is short. It does not take long for the adults to mate and lay the eggs.

4.2 Biology of the larvae

4.2.1 Feeding

During the first hours of their post-embryonic life, the young larvae feed on the mucilaginous substance of the egg mass, then they pierce the envelope to become planktonic and become microdetritivores ingesting small organic particles.

From the second instar, when the larvae are looking for a support, they feed on the detrital film of the bottom, ingesting both dead and living algae as well as inorganic particles [50].

Based upon the feeding mode, larvae can be grouped in six categories: collectors (gatherers and filterers), shredders, scrapers, and predators (engulfers and piercers) [51]. They feed on a great variety of resources including coarse particulated organic matter fragmentation, periphyton, algae, microorganisms; and dissolved organic matter [52].

4.2.2 Locomotion

Larvae display three modes of motility: swimming, crawling and whole-body respiratory undulation. Swimming and respiratory undulation involve the use of metachronal waves of body bending which travel in a head-to-tail direction. Whereas swimming is produced by side-to-side flexures of the whole body, respiratory undulation employs a sinusoidal wave [53].

4.3 Biology of the pupae

The pupal stage links two active stages in the life of insects: the larval and imaginal stages. While the pupal stage of most insects is immobile, the pupae of most Chironomidae are active for a very large part of their existence. In the majority of Chironomidae the pupae move to accomplish three main functions: moulting from the larval cuticle, providing oxygen for respiration, and moving to the surface for emergence of the adult.

4.3.1 Moulting from the larval cuticle

The pupal cephalothorax is formed in the thoracic segments of the larva, and its large volume exercise a pressure on the dorsal suture. Undulations of the abdomen engage the points on the pupal tegument with the larval tegument, leading the pupa forward into the larval thorax. This extra pressure causes the rupture of the suture to push the back of the larval head to the first abdominal tergum, continuous undulations easily release the pupa outward [1].

4.3.2 Locomotion

The pupa displays two swimming modes, somersaulting and eel-like whole-body undulation, the former being principally a brief, escape manoeuvre, the latter being a faster form of locomotion employed to deliver the pupa to the surface prior to adult emergence [53].

The pupae of the majority of chironomidae swim freely, and they are susceptible to flotation and are provided with structures allowing them to adapt to this way of life. Indeed, these pupae have horns with a large plastron to allow them to float. Likewise, in some chironomidae (Pentaneurini) the anal macrosetae are covered with a sticky gelatinous material forming a natatory fringe that allows them to adhere to the substrate and with sudden flexion of the abdomen the pupae move to the surface. Tubicole species only leave their tubes to hatch on the surface of the water, and their anal fringes are therefore not used for swimming.

Figure 21.
A swarm of Chironomus plumosus *over a tree top [55].*

4.4 Biology of the adults

The adults of Chironomidae are aerial. Swarm flight has long been discussed in several studies [54]. Indeed, dense columnar swarms of Chironomidae are often observed, they extend from the tops of trees, the roofs of houses or around lakes (**Figure 21**) [1, 55].

There is a close relationship between the size of the swarm and the number of matings. Indeed, the formation of flight in swarms allows a high rate of mating especially if the population density is low [56]. Thus, the denser swarms attract more females [57].

Most Chironomidae lay their eggs on or near water. The egg laying sites are very varied depending on the species. Indeed, egg masses of Chironomidae have been observed in lakes, rivers, streams [21], rice fields, sea and vegetation in the case of terrestrial species [58].

In general, the eggs are laid on the surface of the water, it can be carried by the wind and the current and travel long distances before the eggs hatch, which contributes significantly to ensure the dissemination of the species despite increased risk of destruction [24].

5. Ecology of Chironomidae

The Chironomidae is the most ubiquitous group of macroinvertebrates and the most abundant in number of species and individuals, moreover they exist in the majority of habitats [2, 29]. Chironomidae invade the sea, nesting at the seashore and living 30 m at the bottom of the ocean [3, 4]. In addition, the larvae of Chironomidae are found in all freshwater and even terrestrial environments [7, 23, 59].

Under certain conditions, such as an extremely low content of dissolved oxygen, the larvae of Chironomidae are the only insects present in the bottom sediments. Extreme variations in temperature, pH, salinity, depth, current and productivity are exploited by the larvae and imagos of some species of Chironomidae [60]. Thus, they are found in the glacial regions of the highest mountains, included at altitudes above 5600 m in the Himalayan massif [61] and remain active at temperatures of - 16°C [62–64]. Some Chironomidae tolerate the high osmotic pressure of coastal waters such as *Clunio marinus* [65–68].

Ecological studies have shown that the distribution of Chironomidae larvae is conditioned by certain environmental factors such as: temperature, depth, type of substrate, trophic level of the environment and chemical factors such as oxygen concentration and pH [69, 70]. Likewise, certain physiological factors intervene in the spatial variations of Chironomidae such as: the physiological adaptations of the species with the physical and chemical conditions of the environment [71].

Chironomidae are primordial candidates for their use in bioindication for several reasons. In fact, they have intimate contact with solid sediment as well as water pores and the surface layer of water for long periods of their life cycle. In addition, they are widely distributed and often the most abundant of all insect species in aquatic ecosystems [72].

The Chironomidae have been used in the classification of aquatic systems according to their eutrophication and the degree of toxicity by heavy metals [70, 73, 74]. In plus, Paleolimnologists have also used Chironomidae as environmental and climatic indicators in retrospective studies of the great lakes to know their life history [75].

6. Conclusion

The Chironomidae is the most ubiquitous group of macroinvertebrates and the most abundant in number of species and individuals. In fact, there are an estimated 6 359 species of Chironomidae worldwide [18]. However, this number is open to discussion since species are still discovered and newly described.

The larvae of Chironomidae are found in all freshwater and even terrestrial environments. It is certain that the clear preference for aquatic habitats no longer needs to be demonstrated: these vary from conventional running waters (torrents, streams, rivers) and stagnant waters (ponds, lakes, rice fields) [70, 76, 77].

Ecological studies have shown that the distribution of Chironomidae larvae is conditioned by certain environmental factors such as: depth, type of substrate, trophic level of the environment and chemical factors such as oxygen concentration [76].

Throughout the Chironomidae family, description of species is traditionally based on adults, and knowledge of immature stages varies across tribes or even across species of the same genus. Indeed, some genera have immature terrestrial stages, other genera have exclusively aquatic larvae. However, many species have unknown immature stages [47].

A specific and generic richness is recorded in the Palearctic and Nearctic regions, this is probably due to the fact that the majority of researchers on Chironomidae are located in these regions. It would be very interesting to broaden the research spectrum of studies on Chironomidae in other regions of the world such as the Mediterranean region in order to better understand the occurrence and distribution of chironomid species and even to discover other species new for the regions and for science.

Author details

Zerguine Karima[1,2]

1 Department of Biology, University 8 Mai 1945 of Guelma, Algeria

2 Laboratory of Biology, Water and Environment, University 8 Mai 1945 of Guelma, Algeria

*Address all correspondence to: karima.zerguine@gmail.com

References

[1] Armitage P. Cranston P.S. & Pinder L.C.V., 1995. Chironomidae - Biology and ecology of non-biting midges. Chapman & Hall, London. 572 pp.

[2] Ferrington L. C. Jr. 2008. Global diversity of non-biting midges (Chironomidae; Insecta-Diptera) in freshwater. Hydrobiologia. **595**, 447-455.

[3] Neumann D., 2003. Adaptations of Chironomids to Intertidal Environments. Annual Review of Entomology .**21**(1):387-414. DOI: 10.1146/annurev.en.21.010176.002131

[4] Brodin Y. & Andersson M.H., 2008. The marine splash midge *Telmatogon japonicus* (Diptera; Chironomidae)—extreme and alien? . Biological Invasions. **11**(6):1311-1317.

[5] Hazra N., Kumar Saha G., MazumdarA., & Chaudhuri P.K., 2011. Records of chironomids of the tribe Pentaneurini (Diptera: Chironomidae) in the Eastern Himalayas of India. Annales- Société Entomologique de France. **47**(3-4):330-339.

[6] Kravtsova L., 2011. Spatial Distribution of Chironomids (Diptera, Chironomidae) in Coastal Zone of South Baikal. Hydrobiol J. **47**(1):3-14. DOI: 10.1615/HydrobJ.v47.i1.10

[7] Delettre Y.R., 2000. Larvae of terrestrial Chironomidae (Insecta: Diptera) colonize the vegetation layer during the rainy season. Pedobiologia. **44,** 622-626.

[8] Lobinske R.J., Cichra C. E. & Ali A., 2002. Predation by bluegill (*Lepomis macrochirus*) on larval Chironomidae in relation to midge sytanding crop in two central Florida lakes. Florida Entomol. **85** (2) : 372-375.

[9] Sharifian M. F., Pasmans F., Adriaensen C., Du Laing G., Janssens G. P., J. , Martel A., 2014. Chironomidae Bloodworms Larvae as Aquatic Amphibian Food. Zoo Biology. **33**(3) : 1-7

[10] Andersen T, Cranston PS,& Epler J. 2013. Chironomidae of the Holarctic region, keys and diagnoses – larvae. Lund (Sweden): Media-Tryck. Supplement 66; p. 563.

[11] Lévêque C., Balian E. V. & Martens K., 2005. An assessment of animal species diversity in continental waters. Hydrobiologia 542, 39-67.

[12] Sublette J. E. & Sublette M. S., 1973. Family Chironomidae. In Delfinado M. & E. D. Hardy (eds), Catalogue of the Diptera of the Oriental Region, Part 1. University Press, Honolulu, Hawaii, 389-422.

[13] Freeman P. & Cranston P. S., 1980. Family Chironomidae. In: Crosskey R. W. (ed.), Catalogue of the Diptera of the Afrotropical Region. British Museum (Natural History), London, England, 175-202.

[14] Cranston P. S. & Martin J. 1989. Family Chironomidae. In: Evenhuis N. L. (ed.), Catalogue of the Diptera of the Australasian and Oceanian Regions. Bishop Museum Press, Honolulu, Hawaii, 252-274.

[15] Sublette J. E., 1992. A Catalog of Nearctic Chironomidae. Annals of the Entomological Society of America 85(4):537-537 . DOI: 10.1093/aesa/85.4.537.

[16] Spies M. & Reiss F. 1996. Catalog and bibliography of Neotropical and Mexican Chironomidae (Insecta, Diptera). Spixiana. **22**, 61-119.

[17] Sæther O. A. & Spies M., 2004. Family Chironomidae (Diptera), on-line @ http://www.faunaeur.org.

[18] Ashe, P.; O'Connor, J.P. A world catalogue of Chironomidae (Diptera). Part 1. Buchonomyiinae, Chilenomyiinae, Podonominae, Aphroteniinae, Tanypodinae, Usambaromyiinae, Diamesinae, Prodiamesinae and Telmatogetoninae; Irish Biogeographical Society & National Museum of Ireland: Dublin, Ireland, 2009; p. 445.

[19] Sæther A. O., 1980. A Glossary of chironomid morphology terminology (Diptera: Chironomidae). Entomol. Scand. Suppl. **14**: 1-51.

[20] Beaumont A. & Cassier P., 1983. Biologie animale des Protozoaires aux Métazoires épithéloneuriens.Bordas, Paris, France. 954 pp.

[21] Williams C. J., 1985. The drift of some chironomid egg masses (Diptera: Chironomidae). Fresh. Biol. **12 (6)** : 573-578.

[22] Kokkinn M.J. & Williams W.D. 1988. Adaptations to life in a hypersaline water-body: Adaptations at the egg and early embryonic stage of *Tanytarsus barbitarsis Freeman.* Aquatic Insects, **4,** 205--214.

[23] Nolte U. 1993. Egg masses of Chironomidae (Diptera). A review, including new observations and a preliminary key. Entomol. Scand. Suppl. **43,** 1-75.

[24] Davies B. R., 1976. Wind distribution of the egg masses of *Chironomus anthracinus* (Zetterstedt) (Diptera: Chironomidae) in a shallow, wind-exposed lake (Loch Leven, Kinross). Fresh. Biol. **6** : 421-424.

[25] Amora G., Hamada N., Fusari L., Andrade V., 2015. An Asiatic Chironomid in Brazil: morphology, DNA barcode and bionomics. ZooKeys. **514**(2):129-144.

[26] Kokkinn M.J., 1990. Is the Rate of Embryonic Development a Predictor of Overall Development Rate in *Tanytarsus barbitarsis* Freeman (Diptera : Chironomidae). Austr. J. Mar. Fresh. Res. **41,** 575-579.

[27] Olafsson J.S., 1992. A comparative study on mouthpart morphology of certain larvae of Chironomini (Diptera: Chironomidae) with reference to the larval feeding habits. J. Zool. London. **228 :** 183-204.

[28] Cranston P. S., 1995a. Introduction. *In*: Armitage, P. D., Cranston, P. S. & Pinder, L. C. V. (Eds). The Chironomidae: biology and ecology of non-biting midges. Chapman & Hall, London, UK, 1-7.

[29] Epler J.H., 2001. Identification manual for the larval Chironomidae (Diptera) of North and South Carolina. A guide to the taxonomy of the midges of the southeastern United States, including Florida. Special Publication SJ2001-SP13. North Carolina Department of Environmental and Natural Resources, Raleigh, NC, and St. Johns River water Management District, Palatka, pp. 526.

[30] www.aquatax.ca/miscdip.html

[31] insects.ummz.lsa.umich.edu/~ethanbr/chiro/

[32] www.waterbugkey.vcsu.edu/php/familylist.php?i...

[33] Langton PH. 1991. A key to pupal exuviae of the west palaearctic Chironomidae. Huntingdon (England): Privately published; p. 150.

[34] Langton P.H. & Wisser H. 2003. Chironomidae exuviae. A key to pupal exuviae of the west Palaearctic region. Amsterdam : Biodiversity Center of ETI.

[35] Pinder L.C.V., 1978. A key to adult males of the British Chironomidae (Diptera). Scientific Publications of the Freshwater Biological Association, **37,** 1-169.

[36] Burtt E.T., Perry R.J.O. & McLachlan, A.J. 1986. Feeding and sexual dimorphism in adult midges (Diptera: Chironomidae). Hol. Ecol. **9** :27-32.

[37] Cranston P. S., 1995b. Systematics. In: Armitage P. D., Cranston P. S. & Pinder L. V. C. (eds), The Chironomidae: Biology and Ecology of Non-biting Midges. Chapman & Hall, London, UK, 31-61, Chapter 3.

[38] Zerguine K., 2010. Contribution à l'étude des Chironomidae (Diptera, Insecta) des mares temporaires de la Numidie Orientale. Aspect de Biologie, Ecologie et Systématique. Doctorat thesis, Badji Mokhtar Annaba University (Algeria). 380p.

[39] Stur E. & Ekrem T., 2020. The Chironomidae (Diptera) of Svalbard and Jan Mayen. Insects, *11*(3), 183

[40] https://alchetron.com/Diamesinae.

[41] http://chirokey.skullisland.info

[42] https://www.flickr.com/photos/130093583@N04/25166332633

[43] Siri A. & Donato M., 2012. Article Two new species of *Podonomus* (Diptera: Chironomidae: Podonominae) of the Brundin's decarthrus group from Ventania system, Argentina. *Zootaxa* . 3548 (3548) : 39-54.

[44] Cranston P.S., 2010. Insect biodiversity and conservation in Australasia. *Annual Review of Entomology*. 55: 55-75. PMID 19961323 DOI: 10.1146/annurev-ento-112408-085348

[45] Pinho L., Carlos S. & Mayumi E., 2018. *Podonomus amarali* n. sp., a new species of the albinervis group (Diptera: Chironomidae: Podonominae) from Atlantic Forest, southern Brazil. Zootaxa . 4402 (3): 542-550.

[46] Marigo T. C., Lamas C.,J.,E., & Fusari L., 2020. A new marine intertidal chironomid from the Brazilian coast (Diptera: Chironomidae: Telmatogetoninae). Zootaxa. 4763(1):117-124

[47] Brundin L., 1983. *Chilenomyia paradoxa* gen. n., sp. n. and Chilenomyiinae, a new subfamily among the Chironomidae (Diptera). Entomol. Scandinav. **14** : 33-45.

[48] Andersen T. & Sæther O.A., 1994. The first record of *Buchonomyia* Fittkau and the subfamily Buchonomyiinae from the New World (Diptera: Chironomidae). *In*: P. Cranston (ed.), Chironomids: from Genes to Ecosystems. CSIRO, Melbourne, pp. 363-367.

[49] Walker I.R. & Mathewes R.W., 1987. Chironomidae (Diptera) and postglacial climate at Marion Lake, British Columbia, Canada. *Quaternary Research*. **27**, 89-102.

[50] Dejoux C., 1981. Chironomides. pp. 583-616. In : Durand J. R. et Lévêque C. (Eds). Flore et faune aquatiques de l'Afrique Sahelo-Soudanienne. Office de Recherche Scientifique et Technique Outre-Mer, Paris, 1023 p

[51] Coffman W.P. & Ferrington, L.C. Jr., 1984. Chironomidae, in *An Introduction to the Aquatic Insects of North America,* 2nd eds, (eds R.W. Merritt and K.W. Cummins), Kendall/Hunt, Dubuque, pp. 551-652.

[52] Henriques-Oliveira A. L., Nessimian J. L. & Dorvillé L. F. M., 2004. Feeding habits of chironomid larvae (Insecta: Diptera) from a stream in the Floresta da Tijuca, Rio de Janeiro, Brazil. *Braz. J. Biol.* **63** (2): 269-281.

[53] Brachenbury J., 2000. Locomotory modes in the larva and pupa of *Chironomus plumosus* (Diptera, Chironomidae). J. Insect. Physiol. **46**: 1517-1527.

[54] Hirabayachi K., Yamamoto M., Takeda M., Hanazato T. & Nakamoto T., 2003. Behavior of Chironomid midges on the bank of lake Suwa, Nagano Prefecture. Pest Control Research. **18** (2) : 91-101.

[55] McLachlan, A. J., 2010. Fluctuating Asymmetry in Flies, What Does it Mean? Symmetry, *2*, 1099-1107.

[56] Kon M., 1987. The mating system of chironomid midges (Diptera: Chironomidae): *a review.* Memoirs of the Faculty of Science, Kyoto University. **12,** 129-134.

[57] Neems R. M., Lazarus J., & Mclachlan A. J., 1992. Swarming behavior in male chironomid midges: A cost-benefit analysis. Behavioral Ecology 3(4):285-290. DOI: 10.1093/beheco/3.4.285.

[58] Frouz J., 1997. The effect of vegetation patterns on oviposition habitat preference : A driving mechanism in terrestrial chironomid (Diptera : Chironomidae). Res. Pop. Ecol. **39** (2) : 207-213.

[59] Avila-Jiménez M.L., Solhøy T., Gwiazdowicz D.J., Fjellberg R.N.E., Dózsa-Farkas K., Monson F., De Smet W. H., Stur E., Ekrem T., & Coulson S.J., 2019. The terrestrial invertebrate fauna of Edgeøya, Svalbard: Arctic landscape community composition reflects biogeography patterns. Polar Biol. **42**(388):3. DOI: 10.1007/s00300-019-02471-x.

[60] McMullin E. R., Bergquist D. C. & Fisher C. R., 2000. Metazoans in Extreme Environments: Adaptations of Hydrothermal Vent and Hydrocarbon Seep Fauna. Grav. Space. Biol. Bul. **13** (2) : 13-23.

[61] Sæther O.A. & Willassen E., 1987. Four new species of *Diamesa Meigen, 1835* (Diptera: Chironomidae) from the glaciers of Nepal. Entomol. Scandina. **29,** 189-203.

[62] Kohshima S., 1984. A novel cold-tolerant insect found in a Himalayan glacier. Nature. **310**, 225-227.

[63] Lencioni V., 2004. Survival strategies of freshwater insects in cold environments. J. Limnol. **63** (1): 45-55.

[64] Bouchard R. W. & Ferrington L. C., 2009. Winter Growth, Development, and Emergence of *Diamesa mendotae* (Diptera: Chironomidae) in Minnesota Streams. Environm. Entomol. 38 **(1):** 250-259.

[65] Neumann D. & Honegger H.W., 1991. Adaptations of the intertidal midge Clunioto arctic conditions. Oecologia, **3,** 1-13.

[66] Frouz J., Matena J. & Ali A., 2003. Survival strategies of Chironomids (Diptera: Chironomidae) living in temporary habitats: a review. Eur. J. Entomol. **100**: 459-465.

[67] Cornette R., Kanamori Y., Watanabe M., Nakahara Y., Gusev O., Mitsumasu K., Kadono-Okuda K., 2010. Identification of anhydrobiosis-related genes from an expressed sequence tag database in the cryptobiotic midge Polypedilum vanderplanki (diptera; chironomidae). J Biol Chem. **285**: 35889-35899.

[68] Gill S.S., Tuteja N., 2010. Reactive oxygen species and antioxidant machinery in abiotic stress tolerance in crop plants. Plant Physiol Biochem. **48**: 909-930. DOI:10.1002/ece3.1833.

[69] Matthews-Bird F., Gosling W., and Coe A., Bush M., Mayle F., Axford Y. and Brooks S., 2016. Environmental controls on the distribution and diversity of lentic Chironomidae (Insecta: Diptera) across an altitudinal gradient in tropical South America. Ecol Evol. **6** : 91-112.

[70] Zerguine K., Bensakhri Z., Bendjeddou D. & Khaladi O., 2018.

Diversity and distribution of Chironomidae (Insecta: Diptera) of the Oued Charef basin, North-Eastern Algeria. Annales de la Société entomologique de France (N.S.). 54 (2) : 141-155DOI: 10.1080/00379271.2018.1435306.

[71] Lindegaard C., 1992. Zoobenthos ecology of Thingvallavatn: vertical distribution, abundance, population dynamics and production. Oikos. **64 :** 257—304.

[72] Warrin E., Din Z. B. & Che Salmah M.R., 2008. Use of Tropical Chironomid Larvae (Diptera: Chironomidae) as Indicator Organism for Aquatic and Sediment Toxicity Tests. Intern. Conf. Environ. Res. Technol. **25:** 14-19.

[73] Di Veroli A., Santoro F., Pallottini M., Selvaggi R., Scardazza F., Cappelletti D. & Goretti E., 2014. Deformities of chironomid larvae and heavy metal pollution: from laboratory to field studies. Chemosphere. **112**:9-17.

[74] Youbi A., Zerguine K., Houilia A., Farfar K., Boudjema S., Berrebbah H., Djebar M. R. & Souiki L., 2020. Potential use of morphological deformities in Chironomus (Diptera: Chironomidae) as a bioindicator of heavy metals pollution in North-East Algeria. Environ Sci Pol Res. **27**:8611-8620.

[75] Belle S., Rius D. and Bichet V., 2017. Combining limnology and paleolimnology to assess the influence of climate change on two lakes in Southern Greenland. Polar Biol. 1-15.

[76] Lobinske R.J., Ali A. & Stout J., 1996. Qualitative and quantitative studies on Chironomidae and selected physico-chemical parameters in two tributaries of the Wekiva river, central Florida. Florida Entomol. **79** (4), 531-542.

[77] Rossaro B., Lencioni V., Boggero A. & Marziali L., 2006. Chironomids from southern alpine running waters : ecology, biogeography. Hydrobiologia. **562 :** 231-246.

Chapter 7

Ecological Aspects of Tabanids (Diptera: Tabanidae) in a Gabonese Cattle Ranch

Ovono Mélodie Audrey Prisca, Mounioko Franck, Zinga Koumba Christophe Roland, Koumba Aubin Armel, Sevidzem Silas Lendzele, Maroundou Audrey Pamela, Acapovi-Yao Géneviève Lydie, Tamesse Joseph Lebel, Simo Gustave, M'batchi Bertrand and Mavoungou Jacques François

Abstract

To embark on an anti-vectorial fight against mechanical vectors of animal trypanosomosis, investigations were undertaken in order to determine the abundance, species diversity and daily activity of tabanids in a cattle ranch in Gabon. The nzi and vavoua traps were used to catch tabanids in three divisions of this ranch. In this study, 616 tabanids were captured: 349 (56.66%) in Division 1, 226 (36.69%) in Division 2 and 41 (6.66%) in Division 3. In the first Division, *T. taeniola* was the most abundant species with an Apparent Density (ADT) of 2.2, followed by *H. pluvialis* (ADT = 1.05). In the second Division, *H. pluvialis* was most abundant with ADT of 1.6, followed by *T. taeniola* (ADT = 0.38). In the last Division, the most abundant species was *H. pluvialis* (ADT = 0.15). Comparing the relative abundance of catches with sites (Divisions), we realized that there was no statistically significant difference in catches with trapping sites. It was noticed that Division 3 recorded the highest diversity index values. We realized that the nzi trap recorded higher tabanid catches than the vavoua trap. The diurnal activity rhythm of the most frequent species encountered slightly differed with prospection sites.

Keywords: tabanids, ecology, abundance, activity, diversity, ranch, Gabon

1. Introduction

Insects are necessary creatures to the living world [1, 2]. Indeed, they are involved in many processes and mechanisms essential for the functioning of ecosystems [3] for example, honey bees, domestic flies and butterflies pollinate our crops [4–6]. Other groups of insects play the role of predators of certain species such as wasps and ladybugs that attack caterpillars and aphids that destroy plants [7]. Similarly, beetles and flies guarantee the decomposition of organic matter, playing a major role in the recycling of essential nutrients to primary producers [7].

However, there are also insect species, including hematophagous dipterans such as tabanids that play a role in the transmission of several pathogens responsible for many diseases [8]. In fact, tabanids are known to be mechanical vectors of trypanosomes including *Trypanosoma vivax*, responsible for African Animal Trypanosomosis (AAT) or Nagana [9, 10]. These insects are capable of colonizing any type of environment including livestock farms [11–15]. Additionally, these insects have negative impacts on the growth and development of livestock [16].

Presently, there are approximately 4400 known species of tabanids [17, 18]. In Gabon, knowledge on tabanids in livestock farms is lacking. However, previous studies conducted by Mavoungou et al. [11] and Zinga et al. [19, 20] at the Ivindo National Park showed that several tabanid species co-existed in sympatry in the different biotopes prospected. In addition, the report of Obame et al. [21] in traditional livestock farms in north Gabon highlighted the presence of blood-feeding flies in this region. Regarding the weakly documented information on tabanids in livestock farms in Gabon, an entomological prospection to determine the abundance and species diversity is indispensible if control operations are to be conducted in this part of the country [22].

In the Nyanga ranch (located in southern Gabon), trypanosomosis is the most common disease of livestock. This disease causes fever, anemia and death in infected animals [23]. In order to establish effective control strategies against tabanids, mechanical vectors of animal trypanosomosis, an entomological study was undertaken to determine their abundance, species diversity and daily activity in the Nyanga cattle ranch.

2. Materials and methods

2.1 Study zone

This study was conducted in the Nyanga Ranch. It is a cattle farm located in the Nyanga province in the south-west of Gabon (**Figure 1**).

The ranch covers an area of 100,000 ha and was created for the breeding of more than 6000 cattle heads [23]. It is made up of three divisions, each of which has its own individual characteristics.

Division 1 has an area of 30,000 ha. It consists of six (6) sections: Bibonga, Galla, Mibamba, Upper Douki, Nyanga and Lower Douki. About 851 animals are reared in this division. In addition, the vegetation of this site is savanna-like.

Division 2 covers nearly 30,000 ha. It is dedicated for breeding, selection and fattening of males. It receives all males (bulls and oxen) after the weaning stage. This division is subdivided into three sections: Kouri, Moukenlengui and Povo. A total of 1754 animals are present in this division. The vegetation of this division consists of savannas and forest galleries.

Division 3 covers an area of 40,000 ha and includes four (4) sections: Yaba, Douli, Voungou and Douxila. This division is used for cattle breeding. Here, there are cows, heifers and calves that are kept until weaning. More than 2500 animals are reared in this division and the vegetation of this division is savanna and forest.

In general, the vegetation of the Nyanga Ranch area is made up of savannas and forest galleries colonized by many plant families including members of the subfamily Gramineae. The area has a rich and diverse fauna including elephants (*Loxodonta africana cyclotis*), buffalos (*Syncerus caffer nanus*), duikers (*Cephalophus* sp.) and bovines (*Bos taurus*, *Bos indicus)* [24]. The Nyanga Ranch has a dense hydrographic network of numerous swamps and rivers such as Nyanga, Douki, Kouri, Mibamba and Douli.

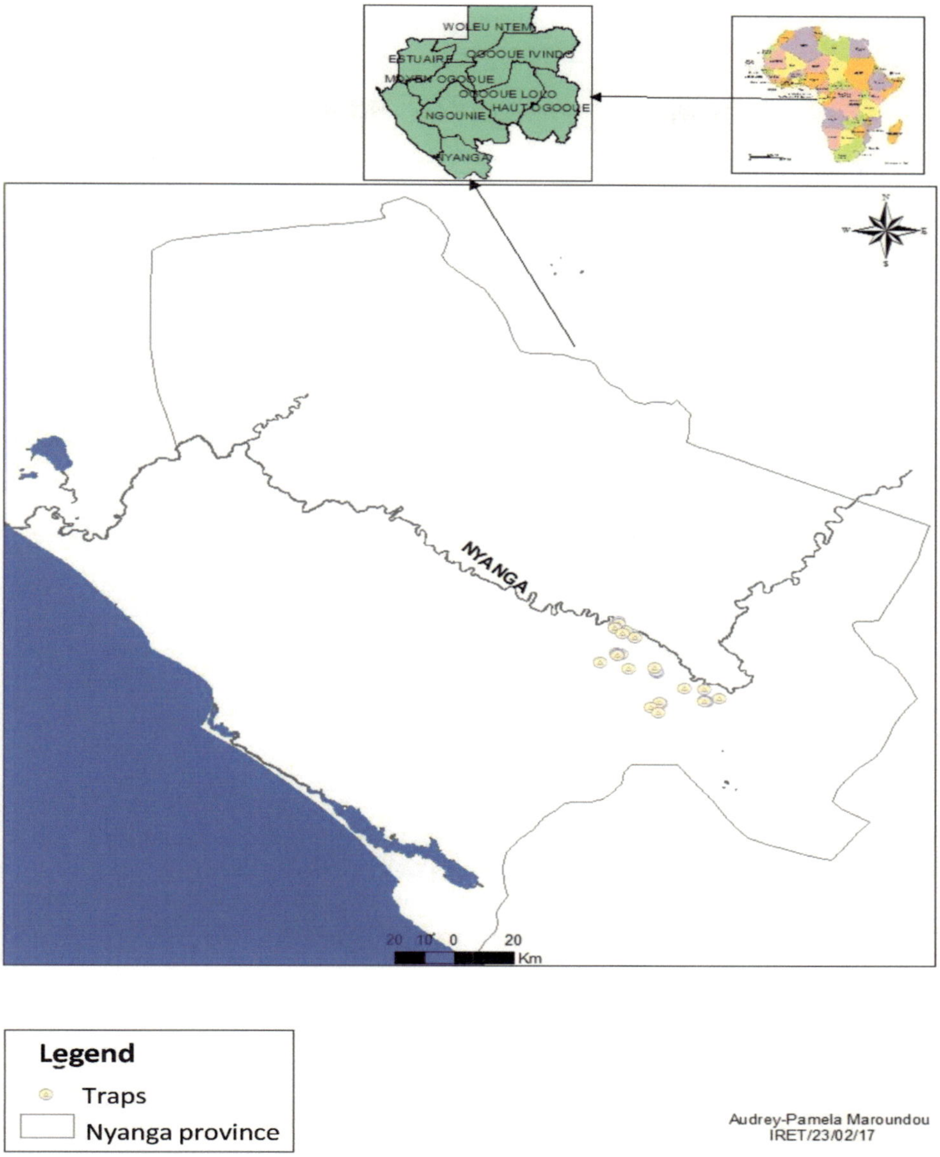

Figure 1.
Map showing the study site and trap positions.

The Nyanga Ranch region has an equatorial climate marked by alternating rainy and dry seasons. The rainy season spans from October to April, while the dry season occurs from May to September [25]. The average annual rainfall is 2000 mm in the North and 1600 mm in the South.

2.2 Capture and identification of tabanids

Tabanids were captured using the vavoua (constructed by Laveissière & Grébau [26] and nzi (constructed by Mihok [27] (**Figure 2a** and **b**) traps. A total of 10 traps were set in each division of the Nyanga Ranch, including 5 traps of each type, spaced approximately 30 m apart. Each of the two traps constituted a capture point in the trapping sites. Flies were collected daily. Trapping was conducted from 9th October to 14th December 2016 with total trapping duration of 60 days.

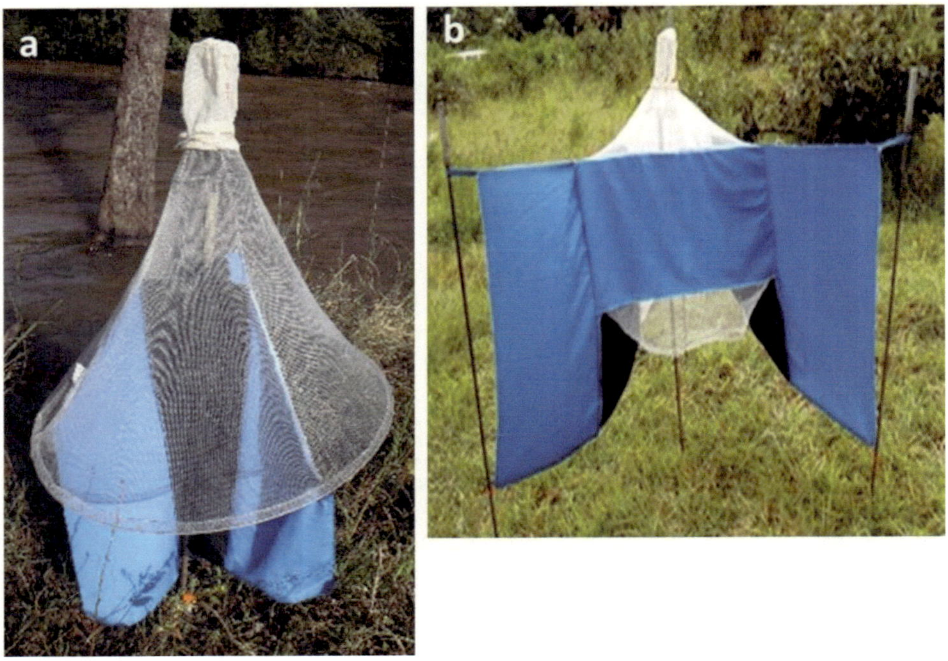

Figure 2.
Trap types a) vavoua trap, b) nzi trap (photos by Sevidzem SL).

In each division, three nzi traps separated by at least 500 m distance were set to evaluate the daily activity of tabanids. The flies were collected systematically every two hours from 8 h to 18 h. The captured flies were put in well labeled vials.

2.3 Fly identification

The identification of tabannids was conducted using the morphological keys of Surcouf and Ricardo [28], Oldroyd [29] and Oldroyd [30].

2.4 Data analysis

The apparent density (ADT) of each species of tabanids was defined as the number of flies caught per trap per days and calculated using the following formula (1):

$$\mathbf{ADT} = \frac{\textbf{Number of tabanid flies captured}}{\textbf{Number of traps} \times \textbf{Number of trapping days}} \tag{1}$$

The biodiversity index of Shannon, which quantifies the heterogeneity of individuals in an environment, was calculated using the following formula (2):

$$\mathrm{H'} = -\Sigma\left(\left(Ni / \mathrm{N}\right) \mathrm{x} \log 2\left(Ni / \mathrm{N}\right)\right) \tag{2}$$

Where
Ni is the number of individuals of a given species.
N, the total number of individuals.

The Simpson index, which is used to determine the probability that two randomly selected individuals in a given milieu are of the same species was calculated using the following formula (3):

$$D = \Sigma Ni(Ni-1)/N(N-1) \tag{3}$$

The Piélou Equitability Index, also known as the Equity Distribution Index, was calculated according to the formula (4):

$$(E) = Ish / \log(S) \tag{4}$$

Where S, is the number of species.

The non-parametric Kruskal-Wallis test was used to compare catches with trapping sites. The statistical test was performed using the XLSTA software version 3.01.19349. The statistical significance level was kept at $p < 0.05$.

3. Results

3.1 Genera composition of tabannids

A total of 616 tabanids divided into 5 genera and 18 species were recorded in this study. The following genera were identified: *Tabanus*, *Haematopota*, *Chrysops*, *Ancala* and *Atylotus* (**Figure 3**).

Figure 3.
Photos showing representative species of each genus identified: a) Tabanus taeniola; b) Ancala species; c) Chrysops longicornis; d) Atylotus agrestis; e) Haematopota species.

3.2 Species composition with trap type

Regarding fly catches with trap types, we noticed that the nzi trap captured more tabanids than the vavoua trap (**Table 1**).

3.3 Proportion of tabanid species

In total, 17 species of tabanids were recorded throughout the survey. The genus *Tabanus* recorded highest species number with frequent species in order of magnitude including *Tabanus taeniola* Palisot de Beauvois, 1807 (33.77%); *Tabanus par* Walker, 1854 (11.36%) and *Tabanus ricardae* Surcouf, 1906 (6.33%). The other species were weakly captured with percentages less than 2% (**Figure 4**). *Ancala fasciata* Fabricius, 1775 (0.49%); *Atylotus agrestis* Wiedemann, 1850 (0.16%); *Chrysops longicornis* Macquart, 1838 (6.33%) and *Haematopota pluvialis* (36.04%) were rare (**Figure 4**).

3.4 The abundance of tabanids with trapping site

Of the 616 tabanids caught in the Nyanga Ranch, 349 (56.66%) came from Division 1, 226 (36.69%) from Division 2 and 41 (6.66%) from Division 3. Of the 8 species of tabanids caught in Division 1, *T. taeniola* (ADT = 2.18) and *H. pluvialis* (ADT = 1.05) were the most abundant. *T. gratus, T par, T ricardae* and *C. longicornis* recorded low ADTs between 0.10 and 0.38 (**Figure 5**).

In Division 2, 13 species were captured and *H. pluvialis* (56%) was the most abundant species with ADT of 1.6 f/t/d, followed by *T. taeniola* (14%) and *T. par* (13%) with ADTs of 0.38 and 0.37 f/t/d respectively. The other 10 species were very weakly represented with ADTs less than 0.2 f/t/d (**Figure 5**).

In Division 3, 10 species were captured and *H. pluvialis* (0.15 f/t/d) and *C. longicornis* (0.15 f/t/d) were the most abundant species with same ADTs. The other

Genus	Species	Nzi	Vavoua
Tabanus (n = 13)	*Tababnus thoracinus*	1	0
	Tabanus dijonctus	1	0
	Tabanus dilitius	1	0
	Tabanus claripes	1	0
	Tabanus laverani	3	0
	Tabanus socius	7	1
	Tabanus taeniola	185	23
	Tabanus obscurehirtus	3	1
	Tabanus ricardae	13	26
	Tabanus par	55	15
	Tabanus gratus	5	3
	Tabanus marmorosus	0	6
	Tabanus sp.	1	0
Ancala (n = 1)	*Ancala fasciata*	2	1
Atylotus (n = 1)	*Atylotus agrestis*	0	1
Haematopota (n = 1)	*Haematopota pluvialis*	46	176
Chrysops (n = 1)	*Chrysops longicornis*	38	1
Total		**362**	**254**

Table 1.
Number of tabanids caught with trap type.

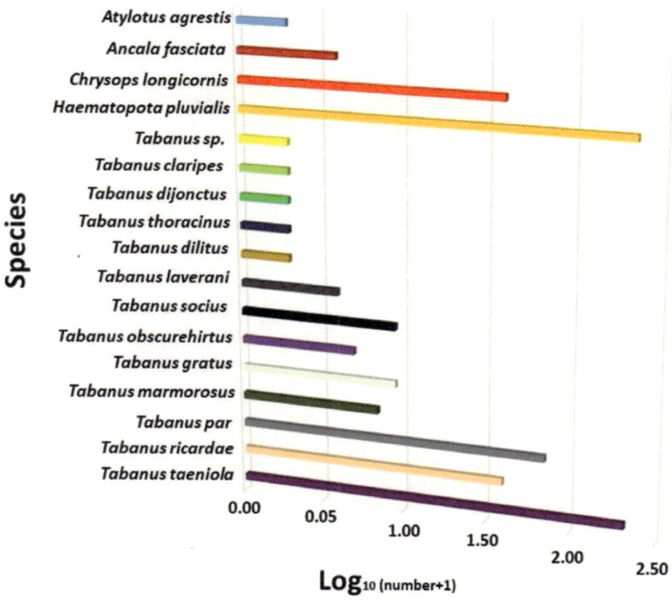

Figure 4.
Frequency of tabanids in the Nyanga ranch.

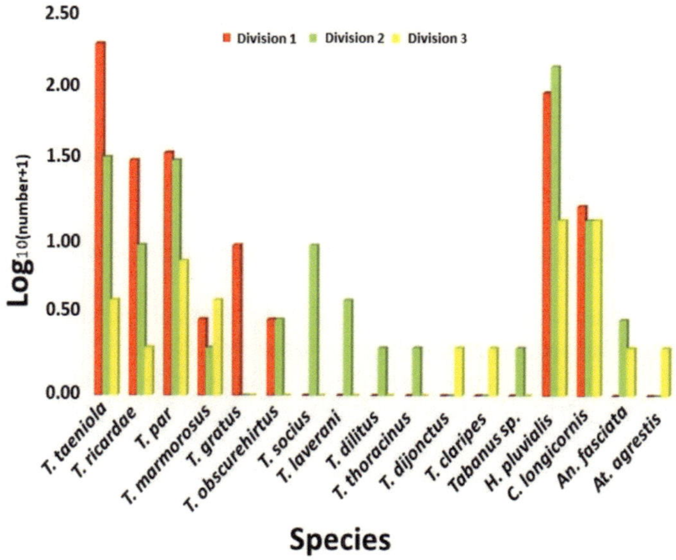

Figure 5.
Distribution of the species of tabanids with respect to the divisions prospected.

species, including *T. taeniola*, *T. marmorosus*, *T. par*, *T claripes*, *T. ricardae*, *T dijonctus*, *Ancala fasciata* and *Atylotus agrestis* had low ADTs below 0.15 f/t/d (**Figure 5**).

3.5 Diversity of tabanids

The results of the ecological indices are presented in **Table 2**. Division 3 showed the highest values in terms of biodiversity index.

The non-parametric Kruskal-Wallis test showed that the species of tabanids caught did not differ statistically with division (**Table 3**).

Ecological indices	Division 1	Division 2	Division 3
Equitability index of Pielou	0.68	0.58	0.78
Shannon index	0.61	0.65	0.78
Simpson index	0.45	0.65	0.79

Table 2.
Ecological diversity indices with respect to the divisions prospected.

Kruskal-Wallis:	
K (observed value)	1.848
K (critical value)	5.991
DDL	2
p-value (bilateral)	0.397
Alpha	0.05

Table 3.
Comparison of tabanids collected from the three divisions using the non-parametric Kruskal-Wallis test.

3.6 Daily activity of the most frequent species of tabanids with respect to prospection divisions

3.6.1 Diurnal activity pattern of H. pluvialis *and* T. taeniola *in division 1*

In division 1, *H. pluvialis* had a bimodal activity peak. This species reached the first activity peak between 10 h and 12 h and a second peak between 16 h and 18 h. Similarly, *T. taeniola* presented a bimodal activity pattern between 10 h and 12 h then between 14 h and 16 h (**Figure 6**).

3.6.2 Diurnal activity pattern of H. pluvialis *and* T. par *in division 2*

The diurnal activity pattern of *H. pluvialis* and *T. par* in Division 2 differed considerably. *H. pluvialis* recorded a unimodal activity peak between 8 h and 10 h, while that of *Tabanus par* occurred throughout the day with three peaks of activity, first between 8 h and 10 h, between 12 h and 14 h and finally between 16 h and 18 h (**Figure 7**).

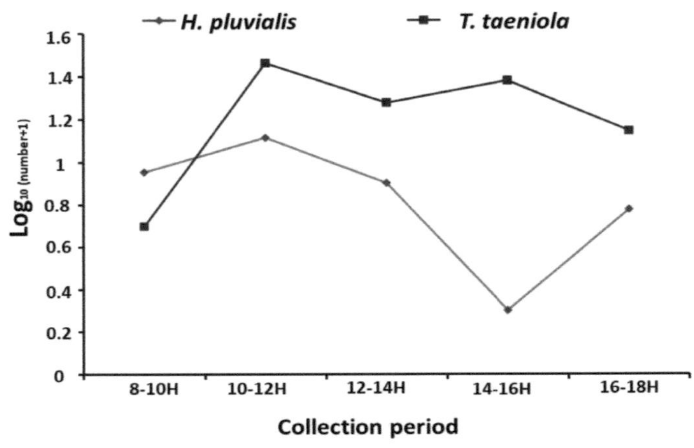

Figure 6.
Daily activity of H. pluvialis *and* T. taeniola *in division 1.*

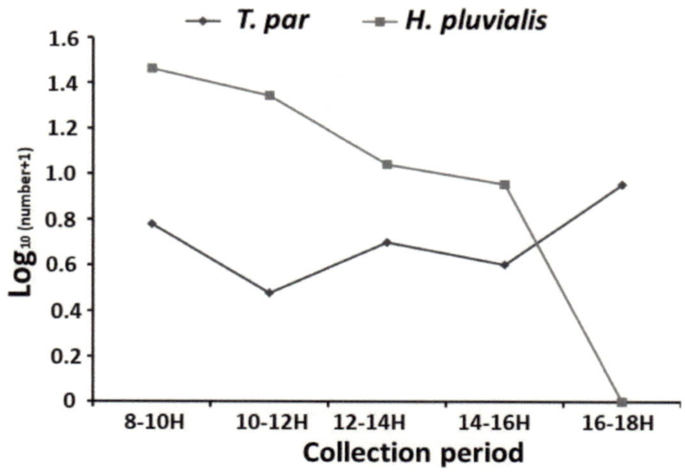

Figure 7.
Daily activity of H. pluvialis *and T .Par in division 2.*

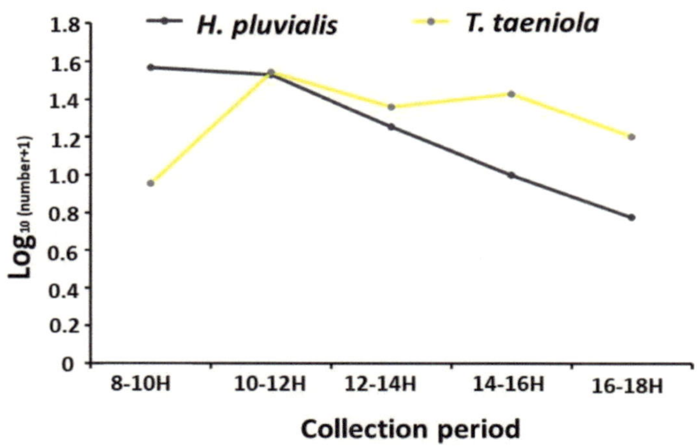

Figure 8.
Daily activity of H. pluvialis *and T. taeniola in division 3.*

3.6.3 Diurnal activity pattern of H. pluvialis *and* T. taeniola *in division 3*

In Division 3, *H. pluvialis* and *T. Taeniola* had an almost similar daily activity patterns. Their activity begins early in the morning and ends up decreasing at the end of the day. Indeed, *H. pluvialis* peak activity occurred between 8 h and 10 h and decreases gradually throughout the day. *T. taeniola* had a bimodal activity pattern, characterized by two activity peaks, one occuring between 10 h and 12 h and the other between 14 h and 16 h (**Figure 8**).

4. Discussion

This study is a preliminary inventory of tabanids, mechanical vectors of AAT in the Nyanga ranch. The results obtained in this study indicates the presence of several tabanid species that live in sympatry in the Nyanga Ranch. Five genera and 18 species of tabanids were identified in this study. We noticed that divisions 1 and 2 were strongly infested by tabanids while division 3 was weakly infested.

This distribution could be explained by the differences in the environmental and microclimatic factors of the divisions which might have created conditions more or less favorable for the development and survival of tabanids. This observation is similar to that made by Mavoungou et al. [11], Zinga et al. [19, 20] and Doumba et al. [31]. These authors observed a heterogeneous distribution of tabanids following the structuring of the prospected milieu. The infestation of an ecosystem by haematophagous flies such as tabanids is defined by the simultaneous presence of many suitable environmental factors, such as temperatures between 15° C and 25° C, good luminosity, high relative humidity and the availability of vertebrate hosts [32–34]. These conditions occurred in Division 1 where maximum catches were made. However, ecological diversity indices especially the Piélou equitability index revealed that Division 3 represented high species diversity.

In addition, some tabanid species including *H. pluvialis*, *T. taeniola*, *T. ricardae*, *T. par*, *T. marmorosus* and *C. longicornis* were captured in the three Divisions but with different ADTs. The other species were identified only in one of the three divisions. This heterogeneous distribution could be partly explained by the ecological requirements of these species and the biotic and abiotic factors of each of the Divisions that favors their development and survival. Moreso, the different vegetation (savanna and gallery forest) composition of the Nyanga Ranch could explain the presence of these species in all the prospected divisions in variable abundances. This finding is similar to that of Acapovi et al. [9] who reported an uneven abundance of tabanids in cattle rearing areas of Côte d'Ivoire.

The dominant species observed in this study were *H. pluvialis*, *T. taeniola* and *T. par*. The abundance of these three species could be explained by their bio-ecology and by the biotic and abiotic factors that exist in this ranch that may favor their development. *H. pluvialis* is known for its preference for forest and wetlands [16, 35]. It is a ubiquitous species that is particularly aggressive to humans and especially animals [11, 13, 15, 36]. The presence of forest and domestic animal species in the ranch could explain the high abundance of *H. pluvialis* in this area. Several studies have reported the occurrence of *T. taeniola* and *T. par* in savanna, forest and livestock areas [11, 14, 16, 31, 37]. In addition, *T. taeniola* is an opportunistic species that has a high adaptive capacity and is therefore found in many environments [9, 11, 30, 38]. Our results on the proportion of tabanid species caught differ from those obtained by Dia et al. [39] in Mauritania, Doutoum et al. [16] in Chad, Mavoungou et al. [11] in Gabon and Acapovi et al. [9] in Côte d'Ivoire. Indeed, in Mauritania Dia et al. [38] obtained 67.5% of *A. agrestis*, followed by 23.4% *T. taeniola* and 9.1% *T. sufis*. In Chad, Doutoum et al. [16] obtained *A. agrestis* (65%), *T. gratus* (22%) and *T. taeniola* (11%) as the dominant species. In Gabon, Mavoungou et al. [11] showed that *Tabanus secedens* (55.2%), *Tabanus obscurehirtus* (13.9%) and *Chrysops dimidiata* (11.2%) were the most important tabanid species. In Côte d'Ivoire, Acapovi et al. [9] reported *Tabanus laverani*, *Chrysops distinctipennis* and *T taeniola* as the most frequent species.

We found that the nzi trap caught more tabanids than the vavoua trap. This observation has been made by several authors who reported that tabanids are mostly attracted to the blue black color of the nzi trap and possibly their size and shape [14, 27, 39, 40].

The results on the daily activity of the various species of tabanids captured portrayed a variation in the number of catches with time of the day. These insects have an activity marked by peaks of abundance observed in the early morning between 8 h and 10 h and at dusk between 16 h and 18 h. In divisions 1 and 3, the activity peak reached between 12 h and 14 h whereas in division 2, the abundance occurred between 8 h and 10 h. These results are similar to those obtained by Mavoungou et al. [41] who showed the importance of hot hours of the day on the abundance of biting flies. Generally speaking, three main species, *H. pluvialis*, *T. taeniola* and *T. par* showed a circadian cycle at twilight with abundance peaks occurring between 8 h and 10 h for *H. pluvialis,* between 16 h and 18 h for *T. par* and 10 h and 12 h as well as 14 h and 16 h for *T taeniola*. These results are identical to those obtained by Auroi [42] who showed that the abundance of tabanids coincides with maximum radiation (the case of *T. taeniola*). However, for *H. pluvialis* and *T. par*, there is a sharp drop in the frequency at 12 h and 16 h. The depression that separates the two peaks of abundance of *H. pluvialis* and *T. par* corresponds to the hottest period of the day when temperatures exceed 27°C. High temperature values are associated with low relative humidity values. These observations corroborate with those made by Gurgenidze [43] who reported plume abundance curves with depressions between 12 h and 15 h when temperatures exceeded 32°C.

5. Conclusion

This study identified 17 species of tabanids with *T. taeniola* recording the highest frequency rate. The daily activity of tabanids differed with species and environment. We found out that Division 3 recorded highest tabanids species diversity. The nzi trap caught more tabanids than the vavoua. A more in-depth study on this taxon is underway to identify the pathogens they harbor.

Acknowledgements

We thank all the workers of the Nyanga Ranch for assisting during field surveys.

Conflict of interest

The authors declare that they have no competing interests.

Author details

Ovono Mélodie Audrey Prisca[1], Mounioko Franck[1], Zinga Koumba Christophe Roland[2], Koumba Aubin Armel[2], Sevidzem Silas Lendzele[6*], Maroundou Audrey Pamela[2], Acapovi-Yao Géneviève Lydie[3], Tamesse Joseph Lebel[4], Simo Gustave[5], M'batchi Bertrand[1] and Mavoungou Jacques François[1,2]

1 Université des Sciences et Techniques de Masuku; BP: 941, Franceville, Gabon

2 Institut de Recherche en Ecologie Tropicale (IRET); BP: 13354, Libreville, Gabon

3 Université Félix Houphouët - Boigny, UFR Biosciences 22; BP: 582, Abidjan 22, Côte d'Ivoire

4 Université de Yaoundé I, Ecole Normale Supérieure, BP: 47 Yaoundé, Cameroun

5 Université de Dschang, Faculté des Sciences, Département de Biochimie, BP: 67 Dschang, Cameroun

6 Ecole Doctorale des Grandes Ecoles, Libreville, Gabon

*Address all correspondence to: sevidzem.lendze@gmail.com

References

[1] Bommarco, R.M., Vaissière, L.B.E. 2012. Insect pollination enhances seed yield, quality, and market value in oilseed rape. *Oecologia*. 169, 1025-1032. Doi: 10.1007/s00442-012-2271-6.

[2] Duvallet, G., Gentile, L. 2012. Protection personnelle antivectorielle. IRD Editions, 25.

[3] Basset, Y., Mavoungou, J.F., Mikissa, J.B., Missa, O., Miller, S., Kitching, R.L., Alonso, A. 2004. Discriminatory power of different arthropods data sets for the biological monitoring of anthropogenic disturbance in tropical forests. Biodiversity and Conservation. 13, 709-732.

[4] Bartomeus, I., Potts, S.G., Steffan-Dewenter, I., Vaissière, B.E., Woyciechowsk, M., Krewenka, K.M., Tscheulin, T., Roberts, S.P.M., Szentgyörgyi, H., Westphal, C., Bommarco, R. 2014. Contribution of insect pollinators to crop yield and quality varies with agricultural intensification. Peer J. 2, 328.

[5] Breeze, T.D., Vaissière, B.E., Bommarco, R., Petanidou, T., Seraphides, N., Harruis, M., Kozák, L., Scheper, J., Biesmeijer, J.C., Kleijn, D., Gyldenkærne, S., Holzschuh, A., Steffan-Dewenter, I., Stout, J., Pärtel, M., Zobel, M., Potts, S.G. 2014. Mismatch of supply and demand of honeybee pollination services across Europe: Implications for service security. PLoS ONE. 9.

[6] INRA. 2015. Les chercheurs volent au secours des abeilles. Dossier Presse; Sciences & Impact, 28.

[7] Brisson, J.D., Mario, F., Bernard, D., Lina, B. 1992. Les insectes prédateurs : Des alliés dans nos jardins. Comment mieux les connaître pour mieux les protéger. Editions Versicolores, La revue québécoise du jardinage, 48.

[8] Krinsky, W.L.1976. Animal disease agents transmitted by horse flies and deerflies (Diptera: Tabanidae). J. Med. Entomol. 13, 225-275.

[9] Acapovi-Yao, G. Cisse, B. Zinga, K.C.R. Mavoungou, J.F. 2016. Infections trypanosomiennes chez les bovins dans des élevages de différents départements en Côte d'Ivoire. Rev. Méd. Vét. 167, 289-295.

[10] Acapovi-Yao, G.L. 2005. Identification et bioécologie des Tabanidés, vecteurs mécaniques potentiels de la transmission de la trypanosomose bovine dans les régions de savanes en Côte-d'Ivoire (Odienné et Korhogo). Thèse de Doctorat, Université de Cocody, Abidjan (Côte-d'Ivoire), 147.

[11] Mavoungou, J.F., Makanga, B., Acapovi-Yao, G., Desquesnes, M., M'batchi, B. 2012. Chorologie des Tabanidae (Diptera) dans la réserve de Biosphère Ipassa-Makokou (GABON) en saison des pluies. Parasite. 19, 165-171.

[12] Baldacchino, F., Muenworn, V., Desquesnes, M., Desoli, F., Charoenviriyaphap, T., Duvallet, G. 2013a. Transmission of pathogens by *Stomoxys* flies (Diptera, Muscidae): a review. Parasite. 26, 1-13.

[13] Baldacchino, F. 2013b. Ecologie des Tabanidae en zones pastorales méditerranéennes et perspectives de lutte. Sciences agricoles. Université Paul Valéry - Montpellier III, 249.

[14] Sevidzem, S.L., Tchawe, R., Zinga-Koumba, R., Mamoudou, A., Ndjonka, D., Mavoungou, J.F. 2019. Insecticide coated screen models reduce insect-vector population in a pasture area in Ngaoundere, Cameroon. Trends Appl. Sci. Res. 14, 80-89.

[15] Lendzele, S.L., Eisenbarth, A., Zinga-Koumba, R.C., Mavoungou, J.F.,

Renz, A. 2019. Aspects of the bionomics of hematophagous symbovine dipterans in a hyperinfested rangeland of Ngaoundere (Adamawa-Cameroon). J. Asia-Pacific Entomol. 22, 1019-1030.

[16] Doutoum, A., Delafosse, A., Elsen, A., Amsler–Delafosse P. 2002. Vecteurs potentiels de *Trypanosoma evansi* chez le dromadaire au Tchad oriental. Rev. Elev. Méd. Vét. Pays Trop. 55, 21-30.

[17] Roskov, Y., Kunze, T., Paglinawan, L., Abucay, L., Orrell, T., Nicolson, D., Culham, A.,Bailly, N., Kirk, P., Bourgoin, T., Baillargeon, G., Decock, W., De Wever, A., Didz ˇiulis, V. 2013. Species 2000 & ITIS Catalogue of Life, 10th December 2013. Species 2000, Naturalis, Leiden, The Netherlands. Digital resource at <www. catalogueoflife.org/col>.

[18] Baldacchino, F., Porciani, A., Bernard, C., Jay-Robert, P. 2013c. Spatial and temporal distribution of Tabanidae in the Pyrenees Mountains: influence of altitude and landscape structure. Bull. Entomol. Res. 104, 1-11.

[19] Zinga-Koumba, C.R., Mbang-Nguema, O.A., Kohagne, T.L., Acapovi-Yao, G.L., Obame, O.K.P., Mutambwe, S., Mavoungou, J.F. 2014a. Contribution à l'évaluation de la diversité des vecteurs biologiques de la Trypanosomose Humaine Africaine et de leur activité journalière dans le Parc National de l'Ivindo (Nord-est Gabon). J. Appl. Biosci. 80, 7060-7070.

[20] Zinga-Koumba, C.R., Mbang-Nguema, O.A., Mavoungou, J.F., Obame Ondo, K.P. 2014b. Ecodistribution des tabanidés, glossines et stomoxes le long d'un transect forêt Primaire-village au Gabon. International J. Biol. Chem. Sci. 8, 167-181.

[21] Obame Ondo, K.P., Zinga Koumba, C.R., Mbang Nguema, O.A., Sembène Mbacké, P., Mavoungou, J.F. 2014. Inventaire des mouches hématophages dans les élevages bovins, ovins et porcins à Oyem (Nord Gabon). Afrique Sci. 10, 373-381.

[22] De La Rocque, S. 2003. Epidémiologie des Trypanosomoses Africaines : Analyse et prévention des risques dans les paysages en transformation. Courrier de l'environnement de l'INRA, n°49 : 80-86.

[23] Maganga, G.D., Mavoungou, J.F., N'dilimabaka, N., Moussadji Kinga, I.C., Mvé-Ondo, B., Mombo, I.M., Ngoubangoye, B., Cossic, B., Mikala Okouyi, C.S., Souza, A., Leroy, E.M., Kumulungui, B., Ollomo, B. 2017. Molecular identification of trypanosome species in trypanotolerant cattle from the south of Gabon. Parasite. 24, 1-7.

[24] Vande Weghe, J.P. 2012. Parc National de Moukalaba Doudou. Agence Nationale des Parcs Nationaux (ANPN), Libreville (Gabon). 296.

[25] Takenoshita, Y.A., Iwata, C.Y., Yamagiwa, J. 2008. Fruit phenology of the great habitat in the Moukalaba-Doudou National Park, Gabon. African Study Monograph Supplementary. 39, 23-39.

[26] Laveissière, C., Grébaut, P. 1990. Recherche sur les pièges à glossines (Diptera : Glossinidae). Mise au point d'un modèle économique : le piège «Vavoua». Trop. Med. Parasitol., 41, 185-192.

[27] Mihok, S. 2002. The development of a multipurpose trap (the Nzi) for tsetse and other biting flies. Bull. Entomol. Res. 92, 385-403.

[28] Surcouf, J., Ricardo, G. 1909. Etude monographique des tabanides d'Afrique. Paris, France, Masson, 292.

[29] Oldroyd, H. 1973. Tabanidae. In: Smith K.G.V. Ed., Insects and other

arthropods of medical importance. London, UK, British Museum (Natural History), 195-202.

[30] Oldroyd, H. 1954. The horse flies (Diptera: Tabanidae) of the Ethiopian region, Vol. II. London, UK, British Museum (Natural History), 341.

[31] Doumba, N.A.G., Zinga Koumba, C.R., Mounioko, F., Maroundou, A.P., Mbang Nguema, A.O., Acapovi-Yao, G.L., M'batchi, B., Mavoungou, J.F. 2016. Contribution à l'étude des stomoxes et tabanides, vecteurs mécaniques des trypanosomes dans la région de Ndendé au sud Gabon. *Entomol Faunistique*. 69, 111-129.

[32] Mounioko, F, Maganga, G.D., Mavoungou, J.F., Zinga, K.C.R., Koumba, A.A., Sevidzem, S.L., Tamesse, J.L., Gustave, S., M'batchi, B. 2018. Molecular Screening of *Trypanosoma* spp. in *Glossina*, *Stomoxys* and Tabanids in the Moukalaba Doudou National Park (South-West, Gabon). World J. Vet Sci. 6, 52-61.

[33] Foil, L.D., Gorham, J.R. 2000. Mechanical transmission of disease agents by arthropods. In Medical Entomology, Eldridge BF, Edman JD (eds). Kluwer Academic Publishers: Dordrecht, the Netherlands, 461-514.

[34] Solano, P., Bouyer, J., Itard, J., Cuisance, D. 2010. The cyclical vectors of trypanosomis. Infectious and Para Diseases of Livestock, 13, 155-183.

[35] Desquesnes, M., Dia, M.L., Acapovi, G.L., Yoni, W., Foil, L., Pin, R. 2005. Les vecteurs mécaniques des trypanosomoses animales : Généralités, morphologie, biologie, impacts et contrôle. Identification des espèces les plus abondantes en Afrique de l'Ouest. Bobo-Dioulasso, Burkina Faso, Editions Cirdes, 68.

[36] Lair X., Livory, A., Sagot, P. 2010. Inventaire préliminaire des taons de la Manche (Diptera Tabanidae). Bulletin Trimestrielle Association Manche-Nature, L'Argiope.70, 1-30.

[37] Acapovi-Yao, G.L. 2005. Identification et bioécologie des Tabanidés, vecteurs mécaniques potentiels de la transmission de la trypanosomose bovine dans les régions de savanes en Côte-d'Ivoire (Odienné et Korhogo). Thèse de Doctorat, Université de Cocody, Abidjan (Côte-d'Ivoire).147.

[38] Dia, M.L., Elsen, P., Cuisance, D., Diop, C., Thiam, A., Chollet, J.Y. 1998. Abundance and seasonnal variations of Tabanids in Southern Trarza (Mauritania). Annals of the New York Academy of Sciences. 849, 456-460.

[39] Sevidzem, S.L., Mavoungou, J.F. 2019. Relative Efficacy of Tsetse Traps and Live Cattle in Estimating the Real Abundance of Blood-Sucking Insects. J. Appl. Sci. 19, 690-700.

[40] Lendzele, S.S., Mamoudou, A., Yao-Acapovi, G.L. 2017. Spatial repartition of tabanids in different zones of north Cameroon. Biodiversity International J. 1.

[41] Mavoungou, J.F. 2007. Ecologie et rôle vecteur des stomoxes (Diptera : Muscidae) au Gabon. Thèse de doctorat. Université Montpellier III PAUL VALERY. 137.

[42] Auroi, C. 1998. Variation quotidienne du nombre horaire de captures de *Haematopota pluvialis* (L.), *Tabanus bromius L.* et *Hybomitra muehlfeldi* (Brauer) (Diptera, Tabanidae), par un piège simulant un hôte. *Bull. Soc. Entomol. Suisse.* 61, 1-11.

[43] Gurgenidze, L.N. 1974. The diel activity of Tabanidae in different landscape areas of Eastern Gruziya, USSR (en Russe). *Parazitologiya.* 8, 249-251.

Chapter 8

Morphological Keys for the Identification of Tunisian *Culicoides* Biting Midges (Diptera: Ceratopogonidae)

Darine Slama, Emna Chaker and Hamouda Babba

Abstract

Culicoides biting midges are tiny blood-feeding insects of several diseases with veterinary and public health significance, including Bluetongue in ruminants, African horse sickness in equids and filarial diseases like Onchocercosis and Mansonellosis affecting various species such as humans. Their identification depends basically on the microscope examination of key morphological characters. Consequently, identification keys are important to any non experiment working with these biting midges. The Tunisian fauna of *Culicoides* biting midges consists of 35 species, whose morphological delineation may be troublesome for non-taxonomists. In response to this situation, and for the first time a key to the adult *Culicoides* species in Tunisia was prepared.

Keywords: *Culicoides*, morphological identification, taxonomy, Tunisia, vectors

1. Introduction

Culicoides Latreille, 1809 (Diptera: Ceratopogonidae) is a genus of biting midges, containing 1368 species divided into numerous subgenera [1]. Indeed, they are vectors of a variety of pathogens, including protozoans [2], filarial parasites [3] such as avian haemosporidians [4] and Tetrapetalonema spp. [5–7]. More than 50 viruses have been isolated from *Culicoides* spp. worldwide [8], such as African horse sicknes virus (AHSV), Bluetongue virus (BTV), Epizootic hemorrhagic virus (EHDV), Schmallenberg virus (SBV), or Oropouche virus (OROV) [9]. These viruses are responsible for outbreaks of non-contagious disease in ruminants, causing severe economic loses [10]. In the Mediterranean basin as well as sub-saharan Africa, the main vector of BTV and AHSV is *Culicoides imicola* [11]. Moroever, other Paleartic *Culicoides* species, within the subgenera Avaritia, such as *C. obsoletus*, *C. scoticus*, *C. dewulfi* and *C. pulicaris* are also known or potential BTV vectors [12, 13]. The recent BTV outbreaks in Tunisia demonstrate how a relatively neglected arthropod vector group can rapidly augment in interest. Thus identification of the vector is an important step in the epidemiology of vector born diseases. For instance, Tunisian fauna of *Culicoides* includes 35 distinct species [14]. Indeed, at the regional level, there are no proper morphological keys to Tunisia *Culicoides* Species. This can cause a major confront in the control effort as the accurate identification of vectors is crucial for vector incrimination. In this context, a morphological identification key was prepared for the adult of the recorded *Culicoides* species in Tunisia as simply as possible, using the most important characters. Hopefully, this will aid public health workers, students and entomologists for rapid and accurate identification of *Culicoides* to the genus and species levels.

2. Methods

2.1 Trapping methods

Culicoides specimens used herein were collected at different locations, in studies conducted in Tunisia. Collection sites were selected based on their characteristics, including presence of animals, type of vegetation, and degree of urbanization. Two models of light traps: home-made miniature using CDC (Centre of Disease Control, Atlanta, USA) and OVI (Onderstepoort Veterinary Institute) were used.

2.2 Samples processing

All insects were collected in a beaker filled with 70% ethanol. Sampled *Culicoides* were separated from other insects and identified according to wing characters using a stereomicroscope. *Culicoides* midges were separated according to sex. Specimens were dissected on a glass slide separating the terminal part of the abdomen, wings and the entire head with a fine needle and mounted in a mix of Balsam Alcohol-Phenol for later identification. Species identification was made according to different morphological keys [15–17].

2.3 Morphological characterization

The morphological characteristics used here were based on original observations and previous usage in the litterature [15, 17, 18] (**Figure 1**). Several morphological characters were examined during the preparation of this morphological identification keys. The following main features were considered: *Head* (Eyes; Antenna (short segments: shape; sensilla coeloconica (Presence); antennal ratio XI/X ratio,

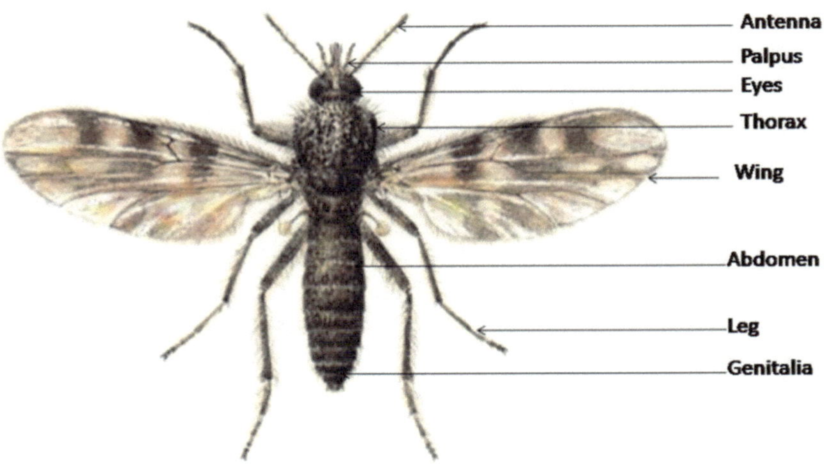

Figure 1.
Adult of Culicoides (Diptera: Ceratopogonidae).

Sub-genera	Species	Wing	Head		Palpus	Spermathecae	Genitalia of male
			Female	Male			
Avaritia	*C. imicola*						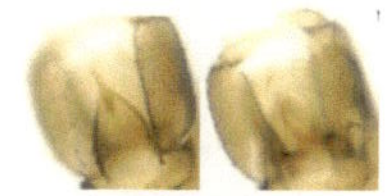
Oecacta	*C. jumineri*		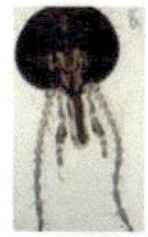		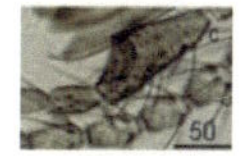		
	*C. gejgelensis or C. cataneii variation**						NA
	C. semimaculatus		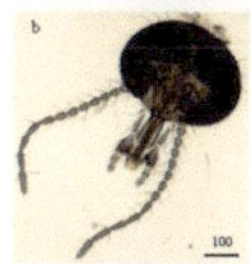	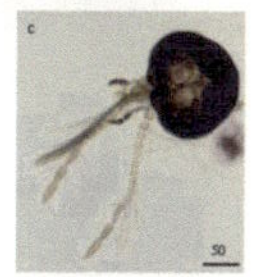			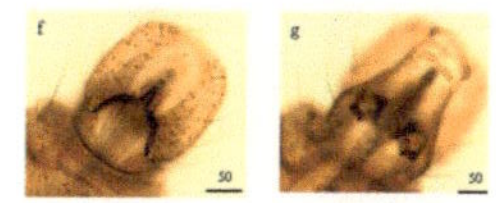
	C. sergenti			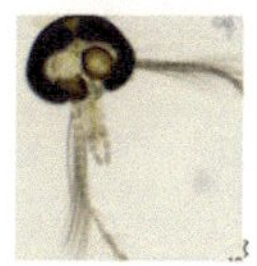			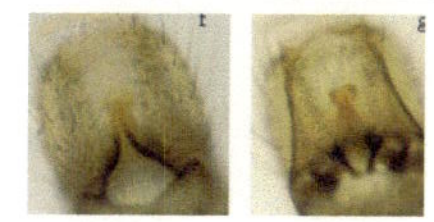
	C.jumineri var	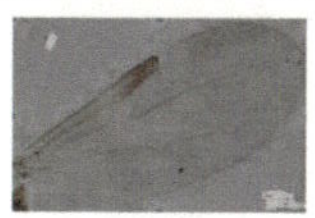	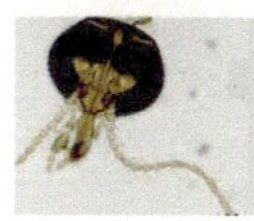	NA	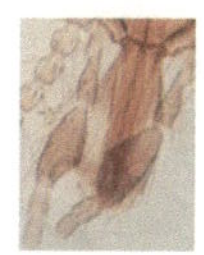		NA

Sub-genera	Species	Wing	Head		Palpus	Spermathecae	Genitalia of male
			Female	Male			
	C.pseudopallidus			NA			NA
	C. cataneii	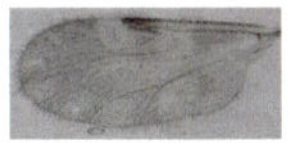		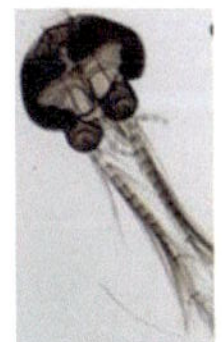		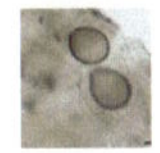	
	C.pseudojumineri						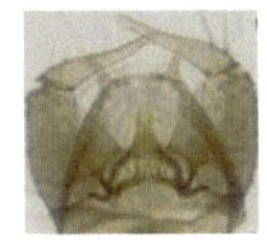
	C.sp near kibunensis				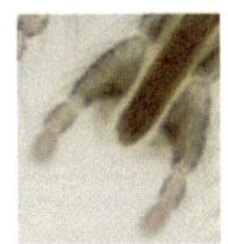		NA
	C.submaritimus : C.maritimus	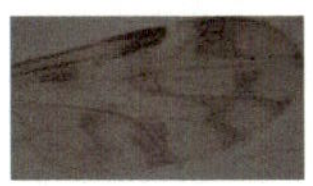			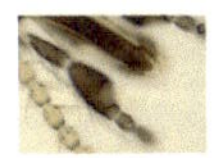		NA

DOI: http://dx.doi.org/10.5772/intechopen.96656

Sub-genera	Species	Wing	Head		Palpus	Spermathecae	Genitalia of male
			Female	Male			
	C.longipennis						NA
	*C. corsicus**			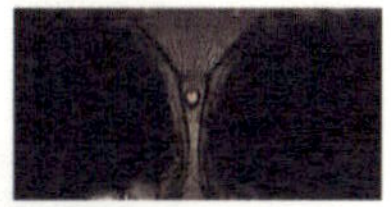			NA
	C. heteroclitus		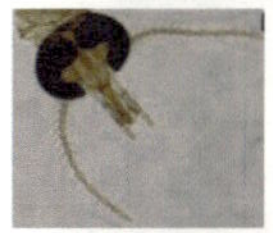			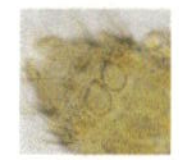	NA
	*C.griseidorsum**			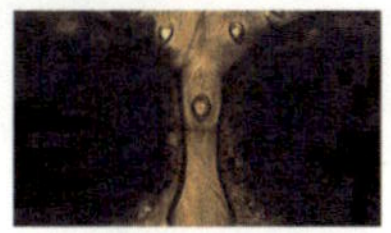			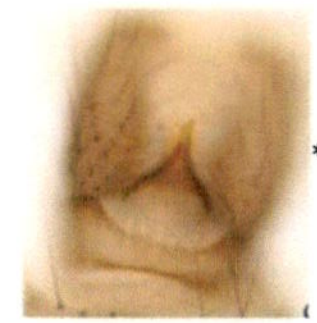 **
	*C.maritimus**			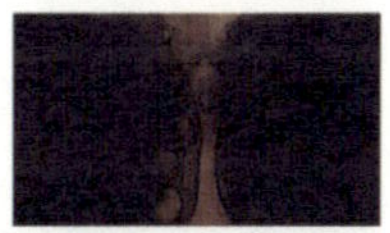			NA
	*C. santonicus**	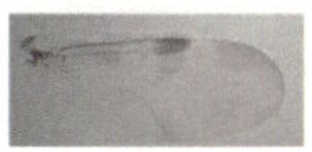					NA

Sub-genera	Species	Wing	Head		Palpus	Spermathecae	Genitalia of male
			Female	Male			
	*C. univittatus**						NA
Synhelea	*C.sahariensis*	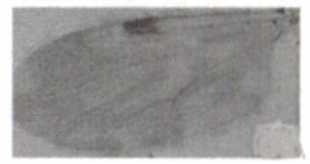					
Culicoides	*C.newsteadi*		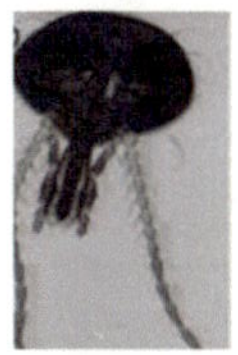				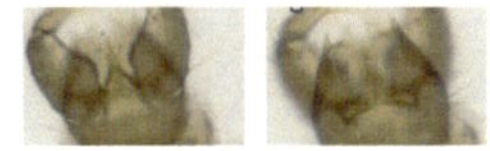
	C.punctatus	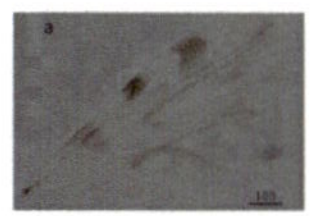	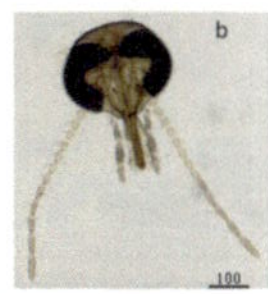	NA	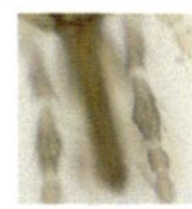	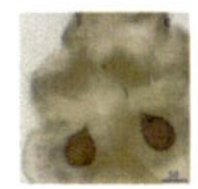	NA
Beltranmyia	*C.circumscriptus*					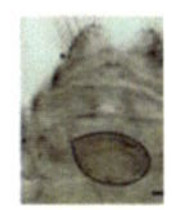	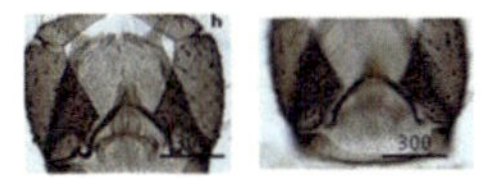

Sub-genera	Species	Wing	Head		Palpus	Spermathecae	Genitalia of male
			Female	Male			
Monoculicoides	*C.puncticollis*				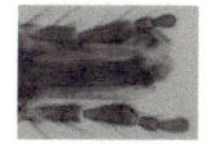		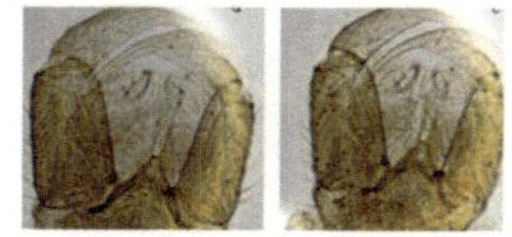
	*C.parroti**	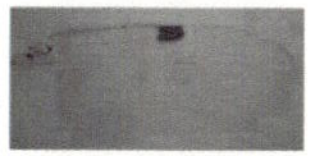			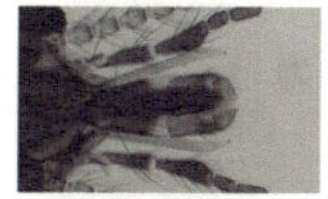	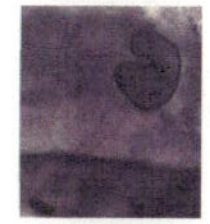	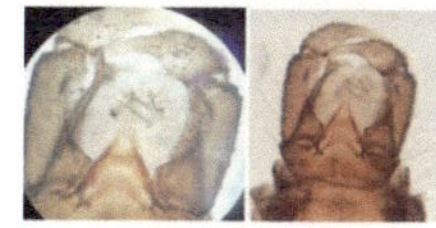
	*C.riethi**	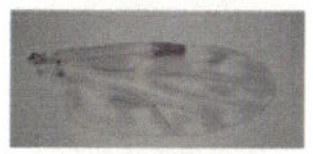	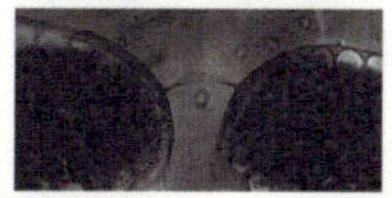				NA
Remmia	*C.kingi*	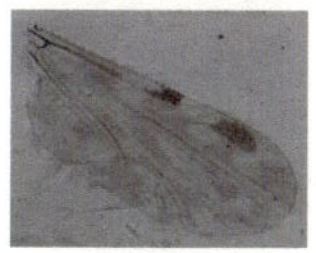			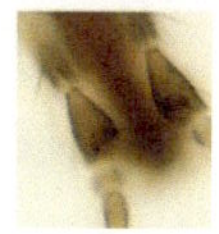		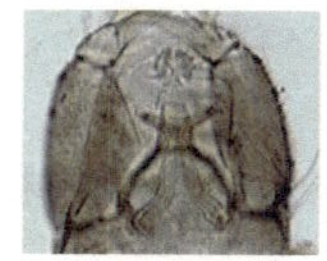
Pontoculicoides	*C.saevus*			NA	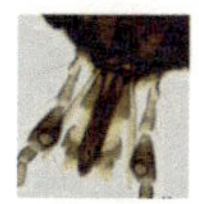		NA
Miscellanoues	*C.paolae*						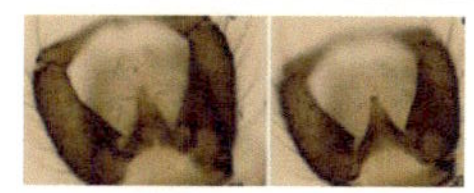

Sub-genera	Species	Wing	Head		Palpus	Spermathecae	Genitalia of male
			Female	Male			
	C.kurensis		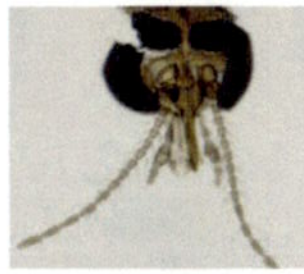		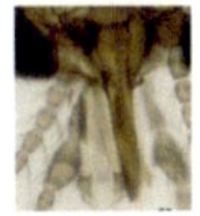		NA
	C.langeroni						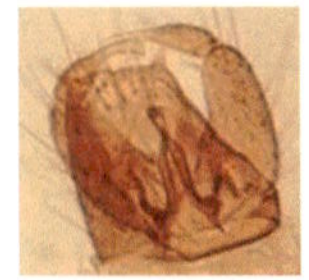
	C.pseudolangeroni				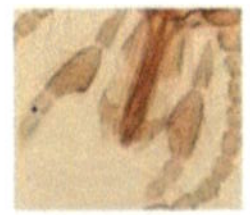	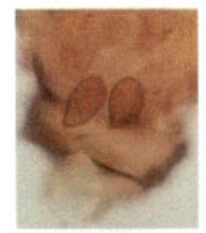	NA
	*C. marcleti**	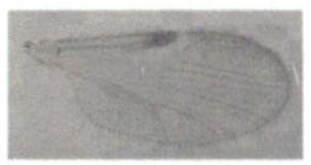	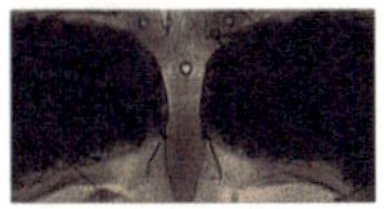		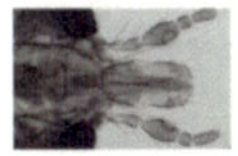		NA
	C. odiatus		NA			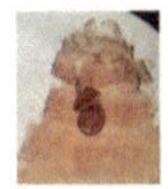	NA
	*C.indistinctus**	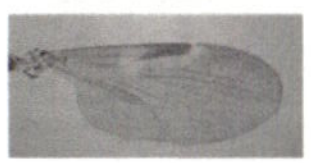	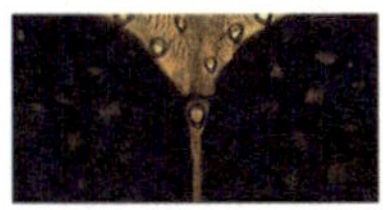		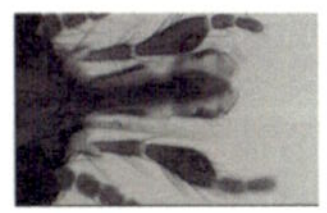		NA

Sub-genera	Species	Wing	Head		Palpus	Spermathecae	Genitalia of male
			Female	Male			
Silvaticulicoides	*C. fascipennis*				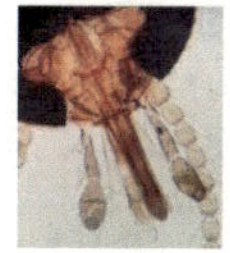		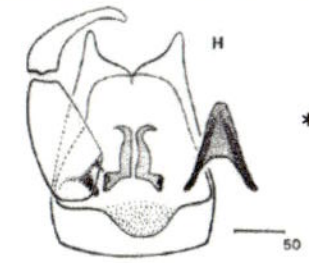 ***
	*C. subfascipennis**	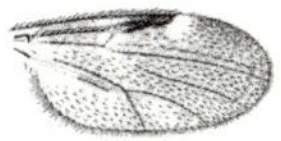	NA			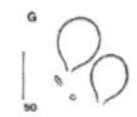	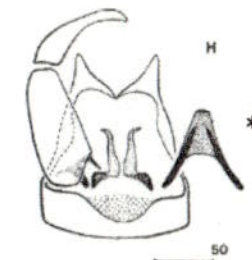 ***

*: *photo of different part (wing, head, 3rd palpal and spermatheca) of* Culicoides *species were taken from Mathieu et al 2012*

**: *photo of genitalia taken from Turgut F, 2018*

***: *photo of drawn part of* Culicoides *were taken from Delecolle JC, 1985*
Other image illustarting the different parts of Culicoides *body in this chapter were taken from the Laboratory of Medical and Molecular Parasitology-Mycology LP3M (code LR12ES08), Department of Clinical Biology B, University of Monastir, Tunisia*

Table 1.
Key to Culicoides *species in Tunisia based on female and male morphology.*

length of segment XI divided by length of segment X; sensilla coeloconica, segments III to VI (Presence); sensilla coeloconica, segments VII to X (Presence); sensilla coeloconica, segments XI to XV (Presence)); *Palpus* (3rd palpal segment (shape, number of sensory pits, single sensory pits (opening versus depth)); *Abdomen*: Spermathecae (spermathecal duct at the end of the sclerotized ring, number, size, shape, abdominal sclerites (presence or absence)); *Wings* pale or dark spots (presence/absence), distribution, size.

Representative *Culicoides* specimens for each species available were selected and relevant characters were image captured with a digital camera through Leica microscopy. For some illustrations of Tunisian *Culicoides* species we used a drawn image used by (18).

3. Results and discussion

The present works illustrated the identification keys to adults of 35 *Culicoides* species (Diptera: Ceratopogonidae) which have been recorded in Tunisia (**Table 1**). Indeed, comprehensive identification keys of genera and species of *Culicoides* were elaborated. The simplified keys allow distinguishing each genus, subgenus and species based on the most important morphological features. Nevertheless, identification keys are fundamental for anyone dealing with insects of medical and veterinary significance, such as *Culicoides* biting midges. They are aimed to provide a guide for those interested to identify field-collected specimens obtained for many purposes and different type of studies. Indeed, identification keys especially those accompagnied by digital photographs and line drawing illustarting taxonomically relevant characters, are useful for species identification of *Culicoides* midges [15]. In Tunisia, even though the first record of *Culicoides* was documented in 1981 [15], field identification of this medically important group of insects has been carried out referring to the basic features and published literature. Indeed, in the case of the Tunisian biting midge fauna, [15] (in French) contain only restricted geographical areas, do not contain the new species records [14, 16, 19]. In the morphological identification keys presented here are accompanied by digital photographs taken from original specimens, which make the keys more user-friendly and facilitating the accurate identification of species.

One of the main limitations of the keys is that the males are incomplete. Therefore, species characterizations were not illustrated in the identification key based on male morphology. Furthermore, some species were not of good quality to take a clear photograph. For this, keys were supplemented with illustrations adapted from published literature [17, 18].

To the author's knowledge, these are the first keys prepared for Tunisian Culicoides fauna which are meant as an aid to the rapid identification.

Key for *Culicoides* (Diptera: Ceratopogonidae) females from Tunisia

1	1 functional spermathecae...	2
	2 functional spermathecae...	5
	3 functional spermathecae...	*C. saevus*
2	Sensilla coeloconica: presence on segments III to XI, presence on segments XII to XIIII...	
	Spermatheca narrow opening...	3
	Sensilla coeloconica: absence on segments XII to XV, presence on segments III, VIII and X...	

	3rd palpal segment slender with slightly swollen, multiple irregular pits...	
	Spermatheca wide opening, oval, elongated, heavily curved in the middle..	**4**
3	Pharynx posterieur armature: presence...............................	
	Sensilla coeloconica: absence on segment XV......................	
	Wing: r-m cross vein, presence on dark spot in the corner with M1 vein..	***C. circumscriptus***
4	Pale wing with only 1 dark spot cover the second radial cell.......	
	Antennal XI/X ratio: inferior or equal to 2..........................	***C. parroti***
	Wing with 1 ore more pale spots.....................................	
	Antennal XI/X ratio: inferior or equal to 2..........................	***C. puncticollis***
5	Eyes: inter-ocular space: joined......................................	
	3rd palpal segment: Multiple irregular pits..........................	
	Wing: 1 ore more pale spots well defined...........................	
	Pharynx posterieur armature: Presence...............................	**6**
	Eyes: Separated..	
	3rd palpal segment: One single sensory pit, triangular, wide opening and shallow pit...	
	Wing: 1 or more pale spot...	
	Pharynx posterieur armature: absence................................	**7**
6	Wing: 2 pale spots fused in the middle basal of M1................	
	Presence of dark spot in r3..	
	Leg: Middle leg: Absence of spines on 4th tarsomere	***C. newsteadi***
	Wing: pale spot in the base of M1	
	Leg: Middle leg: Presence of spines on 4th tarsomere............	***C. punctatus***
	Wing: pale spot over r-m crossvein fused with the m2............	
	Presence of dark spot in r3....................................	
	Leg: Middle leg: Absence of spines on 1st to 3rd tarsomere.....	***C. imicola***
7	Wing: Absence of pale spot (base wing)............................	
	Antenna: antennal XI/X ratio: inferior or equal to 2..............	**8**
	Wing: Presence of pale spot at thebase of the wing................	
	Antenna: I.A: total lengthe of 5 apical segments [11–15]/total length of 8 basal segments [3–10]: superior to 1,12...................................	***C. jumineri***
8	Presence of sensilla coeloconica on the 6th segment...............	**9**
	Absence of sensilla coeloconica on the 6th segment...............	***C. pseudojumineri***
9	Size of 2 spermatheca equal to 50 μ..................................	***C. jumineri var***
	Size of 2 spermatheca superior to 30 μ...............................	***C. heteroclitus***
	Size of 2 spermatheca inferior to 58 μ................................	**10**
10	Presence of sclerotized ring at the end of the spermatheca duct...	***C. gejgelensis or C. cataneii var***
	Absence of sclerotized ring at the end of the spermatheca duct...	**11**
11	Length of palp superior to 185 μ.......................................	***C. pseudopallidus***
	Length of palp superior to 164 μ.......................................	**C. cataneii**

12	Absence of sensilla coeloconica on the segment XV...............	**13**
	Absence of sensilla coeloconica on all segments....................	**16**
13	Length of total antennal segment superior to 517 μ.................	***C.odiatus***
	Length of total antennal segment inferior to 488 μ..................	**14**
14	Palp (3/1 + 2) ratio, length of 3rd palpal segment divided by length of 1st and 2nd palpal segment: inferior to 1, 07............	**15**
15	2 spermathecae: pyriform, size: superior to 58 μ....................	***C. langeroni***
	2 spermathecae ovoid, size: inferior to 50 μ..........................	***C.pseudolangeroni***
	Wing: absence of pale spots in the distal part of r3, m1, m2.......	
	3rd palpal: swollen with narrow opening and deep pit...........	***C. semimaculatus***
	Presence of more than one pale spot in the distal part of r3, m1, m2...	
	3rd palpal segment strongly swollen, with narrow opening single sensory pits and deep pit..	***C. paolae***
	Length of total antenna: inferior to 300 μ...........................	***C. marcleti***
	Length of total antenna: superior to 363 μ...........................	***16***
16	I.A: inferior to 1.03...	***C. kingi***
	I.A superior to 1.20..	***17***
17	Palp (3/1 + 2) ratio: inferior to 1.13...................................	***C. corsicus***
	Palp (3/1 + 2) ratio: superior to 1.26..................................	***18***
18	Presence of chitinized thorn at the beginning of the pharynx posterior..	***C. sahariensis***
	Presence of 1 to 3 thin thorns at the beginning of the pharynx posterior..	***C. longipennis***
	Presence of chitinized thorn at the beginning of the pharynx posterior..	***19***
19	Presence of pale spot r3, m1, m2, anal cell, pale spot in distal part: Presence of 2 pale fused, 2 pale spot separated..............	***C. maritimus***
	Absence of pale spot r3, m1, m2 and anal cell.......................	***20***
20	Absence of pale spots..	
	Presence of sensilla coeloconica on all short segments and absent on all long segments...	***C. sergenti***
	Presence of pale spots..	
	Sensilla coeloconica not present on all short segments.............	***21***
21	Presence of pale spots	
	Sensilla coeloconica on all short segments but often absent on segment X..	
	I.A: superior to 1 μ..	
	3rd palpal segment, triangular, swollen, sensory pit wide opening	***C. sp near kibunensis***
	Presence of pale spot..	
	Presence of sensilla coeloconica on segments VII, VIII, IX and X	
	Presence of sensilla coeloconica on segments XI to XIIII........	
		I.A inferior to 1 μ
	3rd palpal segment, triangular and moderate swollen, single sensory pit, wide opening and swollen pit...........................	***C. kurensis***

Key for *Culicoides* (Diptera: Ceratopogonidae) males from Tunisia.

1	Aedeagus: bifid..	*2*
	Aedeagus: not bifid..	*3*
2	Presence one dark spot on the 2nd rad cell..........................	***C. parroti***
3	Welded paramers with distal part consisting of spoon shaped......	***C. heteroclitus***
	Paramers separated, pointed with 7 to 8 sawteeth on its posterior edge...	***C. sergenti***
	Paramers terminated by a succesion of thorns and presenting a lobe..	***4***
	Paramers with tapered pointed end...................................	*7*
4	Arm of the aedeagus with lateral process............................	*5*
	Arm of aedeagus without lateral process.............................	***6***
5	Body of aedeagus: rectangular, elongated...........................	***C. corsicus***
	Body of aedeagus: short..	***C. marcleti***
6	Body of aedeagus: triangular with truncated end terminating in tiny teeth...	***C. longipennis***
	Body of aedeagus: Triangular with pointed distal end............	***C. sahariensis***
7	Ventral membrane spiculated..	***8***
	Ventral membrane not spiculated......................................	***12***
8	Body of aedeagus: short and large..................................	***9***
	Body of aedeagus: Triangular...	**10**
	Body of aedeagus with rounded apex................................	
	Absence of ventral apodeme...	***C. circumscriptus***
	Body of aedeagus without rounded apex............................	
	Well developped ventral apodeme.....................................	***C. kingi***
10	Body of aedeagus large in the distal part, rounded end	***C. imicola***
	Center of the aedeagus body occuped by a chitinized point directed forward..	***11***
11	Basal quarter of the wing covered by a pale spot.....................	***C. jumineri***
	Slight lightening at the base of the wing..............................	***C. pseudopallidus***
	Wing with pale spots..	**16**
	Absence of pale spots at the base......................................	***17***
12	Body of aedeagus: triangular, wide basal arch andrectangular distal part...	***C. paolae***
	Body of aedeagus: triangular...	***13***
	Body of aedeagus: rectangular..	***15***
13	Body of aedeagus: very little developped...........................	
	Arm of aedeagus: very long..	***C. saevus***
	Arm of aedeagus: well developped....................................	
	Arm of aedeagus: short...	***14***
14	Process: very short, broad-based......................................	***C. langeroni***
	Process: long, with a narrow base.....................................	***C. pseudolangeroni***

15	Body of aedeagus: rectangular, large..................................	*C. cataneii*
	Body of aedeagus: elongated, narrow................................	*C. gejgelensis or C. cataneii var*
16	Single and big pale spot in the basal half of m1......................	*C. punctatus*
	2 or bilobed spot in the basal half of m1..............................	*C. newsteadi*
17	Arms of appendix: horse ear shaped well developped..............	
	Paramers with additional lobe well developped, wide terminal part, with 5 to 6 thiks tips..	*C. semimaculatus*
	Aedeagus without additional appendices on the arms................	
	Paramers without additional lobe......................................	*18*
18	Presence of spot (proximal part of m1)..............................	*C. pseudojumineri*
	Presence of pale spot m4 cut by the edge of the wing	
	Presence of spot (proximal part of m1)..............................	
	Presence of pale spot m4 not cut by the edge of the wing	*C. jumineri var*

4. Conclusion

The simplified keys, would contribute towards improving research capacity among researchers in Tunisia as the identification is a fundamental requirement for anyone dealing with medically important insects.

Acknowledgements

We thank the heads of Hygiene Services of Public Health, all of whom facilitated the field work together, and the health agents who contributed to the achievement of this survey. This study was carried out with the financial support of the Research Lab, Laboratory of Medical and Molecular Parasitology-Mycology LR12ES08, Department of Clinical Biology B, Faculty of Pharmacy, University of Monastir, Monastir, Tunisia.

Conflict of interest

The authors declare no conflict of interest.

Author details

Darine Slama*, Emna Chaker and Hamouda Babba
Laboratory of Medical and Molecular Parasitology-Mycology LP3M (code LR12ES08), Department of Clinical Biology B, Faculty of Pharmacy, University of Monastir, Monastir, Tunisia

*Address all correspondence to: slama.darine@laposte.net

References

[1] Art Borkent. Numbers of Extant and Fossil Species of Ceratopogonidae. 2016; 9(2):118–31.

[2] Tomazatos A, Jöst H, Schulze J, Spînu M, Chanasit JS, Cadar D. Blood - meal analysis of Culicoides (Diptera: Ceratopogonidae) reveals a broad host range and new species records for Romania. Parasit Vectors [Internet]. 2020;1–12. Available from: https://doi.org/10.1186/s13071-020-3938-1

[3] Carpenter, S.; Groschup, M.H.; Garros, C.; Felippe-Bauer, M.L.; Purse BV. Culicoides biting midges, arboviruses and public health in Europe. Antivir Res. 2013;100:102–113.

[4] Chagas, C.R.F.; Bukauskaite, D.; Ilgunas, M.; Iezhova, T.; Valkiunas G. A new blood parasite of leaf warblers: Molecular characterization, phylogenetic relationships, description and identification of vectors. . Parasit Vectors. 2018;11:538.

[5] Yates, J.A.; Lowrie, R.C., Jr.; Eberhard M. Development of Tetrapetalonema llewellyni to the infective stage in *Culicoides hollensis*. J Parasitol. 1982;68:293–296.

[6] Lowrie, R.C., Jr.; Eberhard, M.L.; Orihel TC. Development of Tetrapetalonema marmosetae to the infective stage in *Culicoides hollensis* and *C. furens*. J Parasitol. 1978;64:1003–1007.

[7] Linley JR. Biting midges (Diptera: Ceratopogonidae) as vectors of nonviral animal pathogens. J MedEntomol. 1985; 22:589–599.

[8] Carpenter, S.; Veronesi, E.; Mullens, B.; Venter G. Vector competence of Culicoides for arboviruses: Three major periods of research, their influence on current studies and future directions. Rev Sci Tech. 2015;34:97–112.

[9] Franziska Sick, Martin Beer HK and KW. Culicoides Biting Midges — Underestimated Vectors for Arboviruses of Public Health and Veterinary Importance. Viruses. 2019; 11:376.

[10] Pinior B, Firth CL, Loitsch A, Stockreiter S, Hutter S, Richter V et al. Cost distribution of bluetongue surveillance and vaccination programmes in Austria and Switzerland (2007–2016). Vet Rec. 2018;182:257.

[11] Slama D, Haouas N, Mezhoud H, Babba H, Chaker E. Blood meal analysis of Culicoides (Diptera: Ceratopogonidae) in Central Tunisia. PLoS One. 2015;10(3).

[12] De Liberato C, Scavia G, Lorenzetti R, Scaramozzino P, Amaddeo D, Cardeti G et al. Identification of *Culicoides obsoletus* (Diptera: Ceratopogonidae) as a vector of bluetongue virus in Central Italy. Vet Rec. 2005;156:301–304.

[13] Dallas, J.F., Cruickshank, R.H., Linton, Y.M., Nolan, D.V., Patakakis, M., Braverman Y, Capela, M., Capela, R., Pena, I., Meiswinkel, R., Ortega, M. D., Baylis, M., Mellor P, Mordue AJ. Phylogenetic status and matrilineal structure of the biting midge, Culicoides imicola, in Portugal, Rhodes and Israel . Med Vet Entomol. 2003;17:379–387.

[14] Slama D, Chaker E, Zrelli S, Mathieu B, Delecolle JC et al. Culicoides (Diptera: Ceratopogonidae) Fauna in Central Tunisia. Entomol Ornithol Herpetol. 2016;5:184.

[15] Chaker E. Contribution l'étude des Culicoides (Diptera: Ceratopogonidae) de Tunisie: Systématique, chorologie et écologie. Université Louis Pasteur; 1981.

[16] Chaker E, Sfari M, Rabhi M, Babba H AR. Note faunistique sur les

Culicoides (Diptera: Ceratopogonidae) du gouvernoratt de Monastir (Tunisie). Parasite. 2005;12:359–61.

[17] Mathieu B, Cêtre-Sossah C, Garros C, Chavernac D, Balenghien T, Carpenter S, et al. Development and validation of IIKC: An interactive identification key for Culicoides (Diptera: Ceratopogonidae) females from the Western Palaearctic region. Parasites and Vectors. 2012;5(1):1–11.

[18] Delécolle J. Nouvelle contribution à l'étude systématique et iconographique des espèces du genre Culicoides (Diptera: Ceratopogonidae) du Nord-est de la France. Université Louis Pasteur.; 1985.

[19] Sghaier S, Hammami S, Hammami H, Dkhil A DJ. Entomological surveillance of Culicoides (Diptera: Ceratopogonidae), vector of Bluetongue in Tunisia. Rev Elev Med Vet Pays Trop. 2009;62:2–4.

Section 3

Control of Mosquitos

Chapter 9

Control Strategy for *Aedes aegypti* (Linnaeus, 1762) Population

Taiana Gabriela Barbosa de Souza, Eduardo José de Arruda, Raphael Antônio Borges Gomes, Alex Martins Machado and Antônio Pancrácio de Souza

Abstract

The mosquito *Aedes aegypti* (Diptera: Culicidae), is adapted to different environments, mainly urban ones. They have a high degree of vectorial competence for viral diseases, especially Dengue, the arbovirus with the highest number of cases in the world. The adaptive ability of this insect and the abundance of breeding sites have undermined attempts at population's control, resulting in a high degree of infestation in many regions of the world, resulting in a Dengue endemic. It is important to understand the different nuances of the insect in order to understand the adaptive capacity of this vector, through the knowledge of his behavior, to propose new strategies and engagement of population in proactive actions that allow the population control of this vector, especially in periods of greater proliferation. This chapter discusses population control strategies, in different scenarios and carried out by different researchers, mainly in Brazil.

Keywords: *Aedes* population control, Arbovirus, epidemiological surveillance, mosquitoes control, public health policies

1. Introduction

The *Aedes aegypti* mosquito (Linnaeus, 1762) has always accompanied human migration from his original habitat in northwest Africa, from where the Spanish and Portuguese maritime trade, through the slave trade and the transport of goods from Africa to the new world, allowed the conquest of new areas by him. The phylogenetic analyzes showed that this mosquito has a monophyletic origin from a single strain, domesticated from Africa [1, 2]. This monophyletic group presents some bioecological variations, such as size, color, host preference for blood-feeding, choice of oviposition, larval development, egg dormancy, development time and vector competence (**Figure 1**) [2].

The dissemination of this vector throughout the Brazilian territory was facilitated by the rapid displacement of the rural population to the urban environment without compatible sanitary conditions for the provision of essential services such as sewage and treated water, which contributes to the transmission of diseases, especially those of vector transmission [3–5]. In this context, *A. aegypti* is a vector of several important viral diseases, being able to transmit the Yellow Fever, Dengue, Zika, Chikungunya and Mayaro viruses. Among these arboviruses, dengue stands

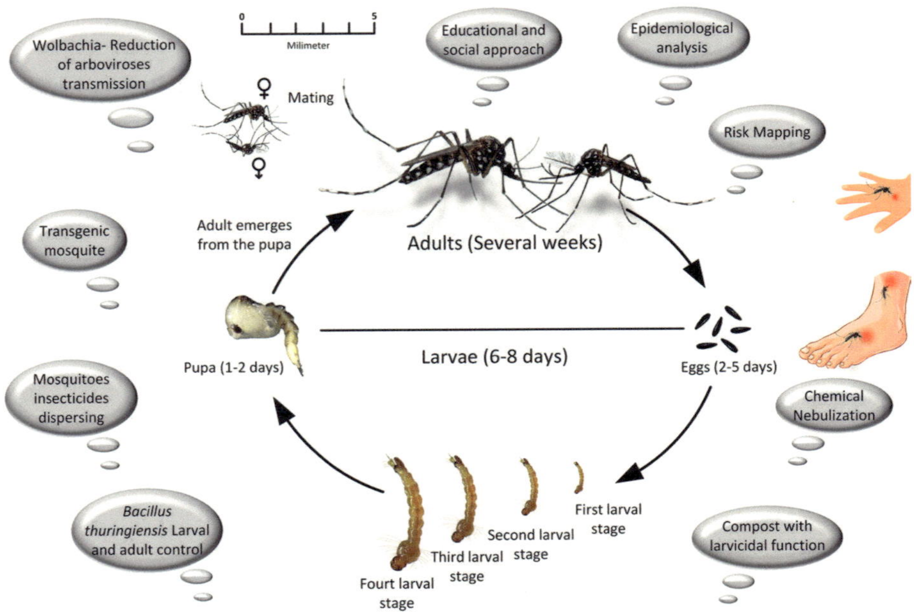

Figure 1.
Aedes aegypti: *Life cycle and control strategies.*

out for the large number of cases, which about 40% of the world population is susceptible to contracting dengue, generating about 500 million cases and 20,000 deaths per year [6].

In Brazil, the first reports of dengue occurred in the late 19th century, in São Paulo (SP), and in the early 20th century, in Niterói (RJ), without laboratory diagnosis. However, in this period, the mosquito was already a problem, but not because of dengue, due to the transmission of yellow fever - which caused numerous outbreaks with a high mortality rate. In 1955, Brazil eradicated *Aedes aegypti* as a result of measures to control yellow fever. The success in eradicating this mosquito in Brazilian territory was achieved with great effort, through awareness campaigns and control of mosquito breeding sites. In addition, the eradication of the *A. aegypti* mosquito was possible at the time because only 20% of the population lived in cities, and the products discarded by them were predominantly organic, not serving as a reservoir for the multiplication of mosquitoes [7].

At the end of the 1960s, the relaxation of the measures adopted led to the reintroduction of the vector into national territory. This period coincided with a period of intense and disorganized urbanization in several Brazilian cities. The industrialization has attracted the rural population to urban areas, leading to an exaggerated urban vegetative growth in a disorganized way and without adequate infrastructure, and many people living in conditions of poverty. The result was the resurgence of the mosquito and the occurrence of epidemic outbreaks of dengue, with the four serotypes, in addition to the proliferation of hemorrhagic dengue (the most serious manifestation of the disease) in all Brazilian states [3, 4, 7, 8].

In 1982 there was a new outbreak of Dengue in Roraima/RR, it was quickly controlled, and considered an epidemic. However, the first time that an epidemic was characterized in Brazil was in 1986–1987 from Nova Iguaçu/RJ, affecting several municipalities and with the largest number of cases in the northeast region. In 1990–1991 a second epidemic occurred, now with more than double the number of cases per 100,000 inhabitants [3, 4, 9, 10]. The Dengue virus was detected with a small outbreak of the disease in Mato Grosso do Sul during the year of 1987. Since then, the epidemics that occur in Brazil are also observed in our state, but especially in the most populous cities, where a moderate increase of cases is visible annually in the months of November and December, the first with the higher number of cases. January to July, coinciding with the increase in temperature, relative humidity and monthly precipitation [11–13].

Due to the growing number of cases, in different cities and regions of the country, the Ministry of Health started to organize itself in national actions and in 1996 launched the *Aedes aegypti* Eradication Program (AaPE). However, the results were not achieved due to the huge socio-environmental changes that Brazil experienced. It must be admitted that the AaPE was important for strengthening national actions to combat the mosquito by increasing the financial resources destined for this purpose, but without achieving a new eradication of the vector [3, 9]. It was evident due to the new outbreak that occurred in 1997–1998, when Brazil faced a new epidemic with more than 500,000 cases occurring mainly in the northeast and southeast regions, with the circulation of serotypes DEN-1 and DEN-2 [3, 4].

In 2001, it was recognized that the mosquito eradication plan was not viable and a new activity plan was launched (called the Dengue Control Action Intensification Plan - DCAIP), with priority for the municipalities that transmit Dengue the most. In order to improve the DCAIP, in 2002, the National Dengue Control Program (NDCP) was launched, with an important change focusing on community mobilization [3, 14].

The implementation of the NDCP has an additional challenge in municipalities on the international frontier due to the movement of people between countries, making it a potential for the occurrence of epidemics; in this sense, dengue is the second most expressive disease in border municipalities. For example, in the municipalities of Corumbá and Ponta Porã (border with Bolivia and Paraguay respectively), the implementation of the NDCP was evaluated and considered partial. There was a demand for improvement, with regard to the training of human resources. It was made like an answer to the inexistence or malfunction of the Municipal Committee for Mobilization, Monitoring and Evaluation of dengue control measures [15].

After a new Dengue epidemic in 2001–2002, the fight against the vector gained the participation of community health agents, who started to carry out preventive and control actions against dengue, according to the Ministry of Health Ordinance No. 44/GM. This initiative was important both for the optimization of resources and the greater involvement of the community in the fight against the mosquito [16–18].

Today, the four Dengue serotypes circulate throughout the country and it is not necessary to introduce a new serotype for new epidemics to occur, since we will always have susceptible people due to births, migration, and time interval between the occurrence of an epidemic with the same serotype [19, 20]. In addition to the number of cases that scare, the economic impact is also impressive and growing [21].

Besides that, the Dengue epidemic with thousands of cases annually across the country, in 2014 the first cases of Chikungunya in Brazil were reported and 8 months later in 2015 the first indigenous cases of Zika virus were also reported, both associated and transmitted by the vector *A. aegypti* [22]. Thus, today, we live in a scenario where there is a triple epidemic associated with this mosquito, where climatic, demographic and social changes were relevant to the current situation, in addition to the intrinsic factors of the pathogens [23].

Some characteristics of epidemics in Brazil are important to highlight, among them the greatest number of cases occur in the first semester and it is important that measures to combat breeding sites and the application of adulticides are carried out as soon as the first cases are detected and not during the period. The epidemic, as transmission between people, occurs quickly and from cases concentrated in a city the number of cases increases rapidly and spreads throughout the city and what is interesting, despite the risk of dengue infection increases with the increase in the population of the mosquito; because infected mosquitoes live less and need to live at least ten more days infected before they are able to transmit the virus [24–26], epidemics are not always related to the size of the mosquito population, but with the susceptibility of the population to the serotypes prevalent in the period [18].

In an attempt to control these important diseases, not only in Brazil but in the world, global strategies have been proposed by WHO. WHO efforts were directed at obtaining reactive responses and intense proactivity for early disease warning systems, use of preventive measures, intense entomological and epidemiological surveillance, search for vaccines and strategies/products for vector control, and reduction of morbidity and mortality. In this context, measures for the population control of the vector through the evaluation of new control agents, mapping of risk zones, storage and logistics, surveillance and early diagnosis capacity, social, educational, and environmental interventions and effective communication between the responsible sectors, can provide efficient ways to control these arboviruses [27–30]. However, despite all efforts made in the last decades, the results were not satisfactory, without being able to effectively control the mosquito population or reduce the incidence of these arboviruses.

We know that viral diseases are complex and require multifaceted responses that involve governments' integrated and global strategy to promote coordinated action between different multisectoral partners with an integrated approach to vector management, sustained control, and measures at all hierarchical levels. The guiding principle should harmonize prevention, entomological and epidemiological surveillance, and efficient case management with existing health systems. The effort must ensure that all strategies be coherent, economically sustainable, and that provide for a reduction in environmental impacts [27–30].

In Brazil, the efforts of the public authorities, especially in relation to the joint action of the federal, state and municipal powers, have been improved, in an attempt to cover all this thinking and strategy guided by WHO, as well as the training of trained human resources different areas, from the rapid diagnosis of the disease, vector control, and determination of risk areas [31]. However, it is still essential to evaluate and create more control measures applied with a robust methodology in order to point out the most efficient practices, worthy of replication and allocation of more resources, within these alternatives.

In this context, our main objective will be to present some alternatives and strategies proposed by researchers, in Brazil and worldwide, to control the vector *Aedes aegypti* and the arboviruses transmitted by it. Still, we have as specific objectives the discussion about the viability of these strategies, as well as a comparison between them, in order to understand and analyze the best methods of population control of the vector.

2. Strategies of *Aedes aegypti* control

Despite efforts to control the population of the mosquito *A. aegypti*, Brazil and other countries suffer annually with epidemics mainly of dengue, with occasional outbreaks of other arboviruses caused by the zika and chikungunya viruses. Uncontrolled urbanization, geographic expansion, vector control programs often lacking adequate resources, and use of inefficient vector control methods, combined with the insect's ability to place its eggs in containers in and around the home has made population control of the vector very difficult. In this scenario, it is necessary to evaluate the strategies adopted so far, and the insertion and evaluation of new techniques in order to identify the most efficient methods in order to allocate the available resources, privileging the most effective actions [32].

It was proposed [32] like a cyclical model of continuous improvement for vector control and, consequently, related viral diseases, with the proposal of an interactive process aiming to improve control programs through the regular and continuous evaluation of methods and techniques used and replacement by better and operationally valid alternatives. The authors propose that proactive control measures should be guided in time and space by epidemiological and entomological data. It is like if the proposed model serves as a catalyst for integrating data on mosquitoes and related arboviruses, filling a gap between control programs, the medical community, and the local government by developing a database that can also supply other activities public health and city planning.

Proactive or prophylactic population control measures for the vector *Aedes aegypti*, such as campaigns to reduce outbreaks, use of insecticide-treated material to protect homes from mosquitoes should have the following characteristics: a) potential for an application not only by control program managers but also by the population in general, b) low cost of execution, c) minimum effort for long-term maintenance. These measures have the advantage of being able to reduce the occurrences of arboviruses related to the controlled vectors [32].

Within that mode, it is very important to encourage community participation, which tends to decrease their concern with these diseases in periods of lower incidence, requiring constant campaigns since the culture and the habit of the population to discard packaging in inappropriate places, in other words, involving the society in campaigns to fight mosquitoes. Thus, the population needs to be informed about the reproductive characteristics of the vector and its biological behavior, in order for the community to be proactively involved, which is an essential condition for success in controlling the vector [32, 33].

Countless campaigns have been carried out, in the most different media to achieve the proactive participation of the population and always targeting the adult population. However, as seen in other awareness campaigns, teaching and understanding the duties of the population, when inculcated in children, has a better effect, by charging children to their parents as well as creating a population more aware of their long-term duties. In this context, it was suggested [34], through the production of informative and interactive booklets, because the education of children generates greater collective awareness in the long run.

On the other hand, there was an interesting study [35] to understand the participation of users in the coproduction of vector control of dengue in Campo Grande - MS, Brazil. It was found that users when included in the relationship with professionals, are able to produce public policy results and benefit from these results. However, the authors still consider that the actions still follow a top-down direction, in the sense that the plan arrives "ready" from the municipality's Health Secretariat, already indicating the actions to be carried out by each member (competencies of agents and actions expected by residents). The autonomy, emancipation, and involvement of managers and authors in the direction of public policy actions have not yet been sufficiently characterized, mainly due to the more effective participation of health agents and users (residents) regarding their responsibility in population control of the vector and reducing the incidence of viral diseases. In this epidemiological scenario, the importance of communication and their effective participation or role can be highlighted, in joint participation in the elaboration of the control plan and its effective application in a continuous and intensive way. In general, experiences show that *A. aegypti* control plans depend on a series of technical data and studies incorporated for decision making and discussion at all levels with ordinary citizens. It is not feasible in order to make vector control decisions.

The epidemiological surveillance system is another sector of strategic importance in the control of vectors, which houses the surveillance of cases of arboviruses (mainly Dengue) and entomological, among others. It is the responsibility of the federal, state, and municipal public authorities that should act in collaborative and synchronous ways. The focus should be on data collection, processing and analysis actions, recommendations for prevention and control measures, as well as the promotion of data collection actions; the processing of collected data; analysis and interpretation of processed data; recommendation of appropriate prevention and control measures. It can promote of the indicated prevention and control actions; evaluation of the effectiveness of the measures adopted and dissemination of relevant information [36, 37].

Incomplete data collection makes it impossible to estimate population risks and allows new epidemics to occur, in addition to reducing the effect of contingency plans and the response capacity of the government to respond satisfactorily in epidemic periods. It is important to carry out periodic assessments of the health surveillance system in general, in order to monitor it efficiently and effectively [38].

It is worth mentioning that the surveillance of cases of dengue and other arboviruses are important to monitor the number of suspected cases to know the time, magnitude and locations of the transmission cases. However, many asymptomatic cases result in silent transmission, so the extent of cases is underestimated. In addition, clinical detection is imprecise and laboratory diagnosis can be time-consuming, which compromises the effectiveness of vector control actions, therefore, interventions to interrupt transmission are impaired [39]. Although, it should be noted that investments in monitoring and case monitoring techniques, as well as the availability of rapid diagnostic tests in health centers, combined with an accurate reporting of each patient's data, can greatly assist in understanding the dynamics of the disease in a municipality, allowing decision-making and effective control methodologies aimed at that specific population.

There is a need for the continuous training of health surveillance professionals, in addition to the constant evaluation of the surveillance system, as well as the carrying out of epidemiological studies that can contribute to interventions in dengue control not only in the state of the study, but across the country [40]. One hundred and thirty-four professionals were interviewed, 70% of whom said they were unaware of the existence of a contingency plan for coping with the dengue epidemic, 59% argued that all suspected dengue cases should be confirmed in the laboratory. Still, one-third of the participants reported difficulty in closing serious cases of dengue [40]. In this context, there is a need for the continuous training of health surveillance professionals, in addition to the constant evaluation of the surveillance system, as well as the carrying out of epidemiological studies that can contribute to interventions in dengue control not only in the state of the study, but across the country.

O vector control must be carried out in response to information from epidemiological surveillance allowing to reduce the transmission force of these viruses, which contributes to better care for people who need treatment. For this, interventions need to be carried out at the beginning of the epidemic peak, at the risk of it being impossible to contain the increase in cases. These interventions require a large amount of human and material resources, intense work, and even the application of insecticides from house to house [39].

Entomological surveillance for the purpose of monitoring to detect the presence and abundance of *A. aegypti*, as well as monitoring resistance to the insecticides used has the advantages of being useful in making decisions about mosquito control interventions, are indicative of the risk of epidemics, and allow the selection of areas and/or periods most critical to the risk of epidemics, in addition to subsidizing the use of more effective insecticides. On the other hand, the disadvantages, in addition to the high cost and low prevalence in mosquitoes. The epidemiological surveillance data are poor indicators for risk of epidemics, because, in addition to the mosquito, the presence of the virus and the population vulnerable or not to the serotype are necessarily circulating, and the vector population may even be under control and still have an epidemic due to the variables related to the virus and the target population [39].

Still, vector control in response to epidemiological data has some important problems, among which we highlight 1) the silent cases that make it difficult to monitor viruses at their onset, especially with new serotypes. 2) interventions often begin with confirmation of laboratory cases, and if this confirmation takes longer, control actions are delayed to prevent an epidemic. 3) the expansion of dengue cases occurs quickly, which makes it difficult to prevent epidemics 4) people move intensely within cities, transporting viruses throughout the city in a short time, making it difficult to monitor and isolate outbreaks of the disease [39].

About the population control of vectors, the most effective methods for the control of mosquitoes that was included a variety of insecticides aimed at controlling adult or immature insects. The implementation of effective control consists of impacting the largest proportion of the vector population. It can be demonstrated that the control strategy must be effective for the high coverage of aquatic mosquito habitats and the reach of winged forms. Among these, it is possible to use methods that use adult mosquitoes to transmit insecticides and other biological products using the behavior for the transfer and dissemination of products between resting and oviposition places in a controlled way to leave residual quantities for extension of population control as a technique of control [41].

In this regard, it is important that *A. aegypti* control activities are adapted to local conditions and their availability of resources to face and control the population. Community engagement is essential, but there are areas where a social organization or local legislation makes such engagement difficult. The application of adulticides by means of adapted vehicles is considered inefficient, often used when there are no other viable alternatives [39].

An important indicator of the mosquito infestation index in a given area or region is RISAa (Rapid Index Survey of *Aedes aegypti*), which is a control method that aggregates the building, *Breteau*, and container indices used to calculate larval density, being important for making decisions about adult mosquito control. The building and *Breteau* indices are more robust than the container index and less sensitive to show sample variations for pupae and adults indices. Pupa rates per person and per household are less robust than pupae per hectare; Similar results were found with the adult mosquito indices. By this method, each city is divided into blocks and a number calculated according to the degree of the infestation, with a satisfactory one, from 1 to 3.9 alertness and above 3.9 with the risk of the epidemic [42]. The RISAa method is unreliable for some authors because even with low rates, dengue epidemics can occur [43, 44].

The traditional method of dividing the city into blocks does not allow the visualization of the city in continuity precisely by dividing the method, which obviously the mosquito is not limited to these blocks in its locomotion in the environment. Thus, methods that can evaluate the blocks in greater detail, detecting points of greater infestation that were not possible for observation by RISAa are possible by using the Gaussian Kernel method. This method, although it has a certain subjectivity which requires knowledge from the researcher, allows a quick and easy view of the risk sites without the barriers imposed by the administrative political organization [45, 46].

It was performed an excellent non-systematic literature review [47] regarding *A. aegypti* population control strategies. The control strategies considered, such as selective monitoring of infestation, social measures, dispersion of insecticides, new biological control agents, and molecular techniques. The authors considered the integrated use of different compatible and effective techniques according to the region to be possible for the possible reduction of the vector and the related arboviruses. The authors also consider that in the case of technologies in development, they still require evaluation as to their effectiveness, feasibility, and costs for their use in conjunction with other techniques already recommended by the National Program for Dengue Control (NPDC).

However, new strategies have been proposed to control the mosquito population and reduce the incidence of diseases [48]. Among these new techniques, we have the genetic modification of mosquitoes in the laboratory, which, when released to the environment, spread the modified genes to the native population, leading to a decrease in this population or its extermination.

The use of insects inoculated with *Wolbachia* could be a step forward for vector and disease control for longer periods in endemic areas around the world. Different studies have shown that the most efficient approach to control transmission can be obtained from the finding that about 60% of insect species carry *Wolbachia pipientis*, however, it is important to note that this bacterium does not naturally infect the mosquito *A. aegypti*, having to be infected in a laboratory environment, generating some production costs for these modified insects. These results show that the technique using *Wolbachia*, which has been in development since the 1990s, could be an interesting option for the reduction of mosquito-borne diseases [49, 50].

In the same vein, the use of *Bacillus thuringiensis* for the control of larvae and mosquitoes has stood out among the various strategies that make up integrated management programs, being more advantageous in relation to chemical insecticides, both in cost and in their action. The insecticidal activity is due to the toxic proteinases present in the bacteria, which when reaching the insects' intestines unfold the protoxins creating pores that interfere with the ion transport system through the tissue membrane, resulting in insect death. Efficiency studies of this methodology affirm an efficiency of more than 70% for a period of 40 days after exposure [51].

Despite the different strategies mentioned here, summarized in **Table 1** and **Figure 1**, with their potential effectiveness, it is necessary to continue the search for methodologies to control arboviruses and their vector mosquitoes, requiring the development of diversified research involving both ecological aspects, behavior, and population biology. It was a way to increase the success of the control methods used, as well as, promoting conditions for the implementation of new control tools, including knowledge, education, and cultural habits of the population.

The authors consider that partnerships between research centers (universities and institutes) and the government are important parts for the elaboration of strategies that are more appropriate to the location with resources that can be made available for the implementation of these strategies in a pilot plan like a way to evaluate the results by epidemiological and entomological criteria [39] on a continuous basis and with reassessments of the effect achieved in the programs and strategies employed. In Brazil, municipal and state committees have been organized with the presence of members from universities and research institutes, education departments, the legislature, the armed forces, as well as others leaders of organized civil society to better articulate the actions to combat this vector. These partnerships allow articulated actions in large-scale and the solution of problems related to mosquito control in a holistic way, involving different public and private sectors for intelligent decision making.

Strategies	Advantages	Disadvantages
Educational and social approach	Involvement and awareness of the population in home control of mosquito breeding sites.	It depends on the involvement of the population and the various sectors of society. Decrease in engagement during the period with the least number of cases.
Epidemiological analysis and risk mapping	It allows the precise analysis of risk regions allowing the correct targeting of resources.	Despite showing the critical regions, it is necessary to be allied with other technologies to be satisfactory.
Intra and extra-home nebulization.	It has spatial coverage and reduces disease transmission at the time of the outbreak.	Can promote selection of resistant populations insecticide; demand application agents trained; little adulticide availability.
Natural/synthetic compounds to larvicidal function	Alternative and safer products when compared to chemical insecticides. Synergistic compounds can increase the larvicidal function of natural or synthetic compounds.	Need of cost-effectiveness studies compared to chemical insecticide.
Transgenic mosquitoes	It leads to a reduction in the life span of mosquitoes; decreases infestation of mosquitoes; and dispenses with the use of radiation.	There is a need to use mosquito sexing technologies; depends on the protocol of release; requires constant production and release mosquitoes.
Wolbachia reduction of arbovirus transmission	Use of microorganism that causes a natural, self-sustaining infection, does not use insecticides and radiation.	Climatic differences, mosquito release protocols, level of urbanization and human density can limit the potential functions.
Mosquitoes insecticide dispersing	Use of larvicide already available and attested agents familiar with the type of trap used; mosquitoes take larvicides for breeding, eliminating them.	Promotes selection of resistant mosquito populations, requires insecticides with ideal concentration in small particles.

Table 1.
Summarization of new Aedes aegypti *mosquito control strategies.*

Finally, the authors conclude that the erradication of *A. aegypti* by top-down approaches how it already happened in Brazil some decades ago it is impossible today because with the rapid immigration of people from rural to urban areas without minimum sanitary infrastructure promoted outbreaks each more commun. The ocurrence of four dengue serotypes allied with constant number of susceptible people to arbovirus due to migration and births during the time interval between the occurrence of an epidemic with the same serotype are the main conditions to Brazil be a favorable local to epidemies frequently. A sustanaible control of dengue and other arboviruses relationed to *A. aegypti* must have the following steps: 1) A continuous improvement of survaillance system, 2) a good control plan linked to epidemiological survaillance, 3) a selection and continuous evaluation of control strategies to *A. aegypti* adapted to each local, 4) an excellent interaction among different social actors to define, apply, evaluated and improve better solutions to each local, 5) use of compatible control strategies among each other, and 6) effort to maintain always the engagement of local community.

3. Conclusion

The *Aedes aegypti* mosquito (Diptera: *Culicidae*) is adapted to the urban environment due to the large supply of artificial breeding sites which result in unsuccessful population control with a high degree of arbovirus spread and infestation in different regions of the world. In this scenario, the experiences over decades of population control, requires understanding about the reproductive success of the species and the adaptability of the vectors of the species *Aedes spp.* It is important to understand the nuances and details of the habitats, behaviors, habits and the ecology of the insect, and to plan the development of new products and strategies that are compatible with each other, that enhance the biological activity and scope of the control, that stimulate the population's adhesion proactive actions before insect proliferation and infestation and reduce possible environmental impacts.

Despite the proposal for different integrated population control strategies, such as breeding elimination, combined chemical control, genetic modification of mosquito populations, chemical control is still one of the most used tools for containing the insect and reducing impacts on public health. However, population control is still unsatisfactory due to the behavior, resistance to insecticides and survival strategies and adaptability, besides high fertility rate of the insect, which despite an apparent fragility, has overcome the restrictions and conditions imposed on its population control. Integrated preventive control is appropriate as long as it considers aspects of the behavior and habits of the target insect, residual and comprehensive activity, and that it can reduce the viability of breeding sites and eliminate, preferably, the egg banks present in the breeding sites. In addition, the voluntary service of the population in the control of the vector in homes and public spaces is essential, there is adequate sanitation infrastructure, health education for the community, stimulating the community's adherence to the vector's domestic control due to its anthropophilic habit. The reports show that immediate successes are not lasting and that all population control strategies, in isolation, present inefficiencies in the medium and long term or even that they present inconclusive results from the analysis of reduction of *Aedes* spp. infestation and disease incidence.

This chapter discusses new products, strategies and proposals for the population control of *Aedes* spp., considering different scenarios and using content and perceptions of experimental results made available by different researchers, mainly in Brazil. Careful analysis of the literature showed that most of the population control failures are probably due to the use of inadequate products to which there is resistance acquired by the insect and\or poorly planned strategies in population prevention or control or even inadequacy detection aspects, quantification of risk analysis that should be used in the control of vectors. The use of products, the strategies used and the application and\or environmental conditions are not periodically reviewed and compromise the effectiveness of the application of (bio) actives or insecticides that are used, in addition to the contribution of environmental and climatic factors and\or restrictions imposed or few resources made available for combat that severely affect control due to the low insecticidal activity of products and applications that are ineffective for resistant urban populations. Insecticidal or control products do not have a broad spectrum of activity or comprehensive control for the different forms of the insect, from egg to winged insect. The products do not have multifunctionality or are not yet presented in the form of intelligent controlled release of (bio) assets to obtain a more prolonged control of activity in breeding sites. Thus, all these control factors combined and\or applied in an inadequate manner and\or severely affect the effectiveness of the vector population control and allow the continuity of the reproduction and transmission of diseases by arbovirus, and, which still has a potential of growth for new diseases and the spread of other arboviruses due to the insect's competence and vector potential.

4. Future perspectives

Based on the contents, reports and data presented, it is proposed that new perspectives of population control should consider an integration between preventive and corrective forms, if necessary, based on the combination of different products and\or techniques that are compatible and synergistic in the application, due to the acquired resistance of the insect and\or the use of control strategies and\ or applications of these products that are available in the regions. Still, it is important that applications of a single type of product or techniques are never carried out in isolation, which result in inefficient and non-lasting treatments, especially in conditions of high insect infestation mainly without considering environmental, climatic factors or local control peculiarities.

In this perspective, the need for a multidisciplinary approach is reinforced with the use of new technologies and products and\or combinations of different potentialized products in the form of smart devices with slow release for lasting (residual) control of the insect population, especially in breeding grounds. We can highlight as highly promising the strategies as follow: 1) an eco-bio-social approach, by focusing on social participation in insect control, in addition to compatibility with other strategies, in addition to dispensing with the use of insecticides, 2) risk mapping, by increased control accuracy, 3) *Wolbachia*, for self-sustainability and efficiency, 4) insecticide-dispersing mosquitoes, for optimization of human resources and compatibility with other strategies, in addition to combinations of techniques that can increase population control. These strategies stand out because they maintain two crucial pillars in the control of this vector: social participation and compatibility with other control strategies.

Author details

Taiana Gabriela Barbosa de Souza[1]*, Eduardo José de Arruda[2], Raphael Antônio Borges Gomes[3], Alex Martins Machado[1] and Antônio Pancrácio de Souza[4]

1 Federal University of Mato Grosso do Sul, Três Lagoas, MS, Brazil

2 Federal University of Grande Dourados, Dourados, MS, Brazil

3 Federal University of Ouro Preto, Ouro Preto, MG, Brazil

4 Federal University of Mato Grosso do Sul, Campo Grande, MS, Brazil

*Address all correspondence to: taiana.souza@ufms.br

References

[1] Brown JE, Evans B, Zheng W, Abas V, Barrera-Martinez L, Egizi A, Zhao H, Caccone A, Powell JR. Human impacts have shaped historical and recent evolution in *Aedes aegypti*, the dengue and yellow fever mosquito. Evolution. 2014:68(2):514-525. DOI: 10.1111/evo.12281.

[2] Salles TS, Sá-Guimarães TE, Alvarenga ESL, Guimarães-Ribeiro V, Meneses MDF, Castro-Salles PF, dos Santos CR, Melo ACA, Soares MR, Ferreira DF, Moreira MF. History, epidemiology and diagnostics of dengue in the American and Brazilian contexts: a review. Parasites & Vectors. 2018:11(1):264. DOI: 10.1186/s13071-018-2830-8.

[3] Terra MR, Silva RS, Pereira MGN, Lima AF. *Aedes aegypti* e as arboviroses emergentes no Brasil. Revista Uningá Review. 2017:30: 3. Available from: http://revista.uninga.br/index.php/uningareviews/article/view/2028.

[4] Lopes TRR, Silva CS, Pastor AF, Silva Junior JVJ. Dengue in Brazil in 2017: what happened? Revista Instituto Medicina Tropical de São Paulo. 2018:60. DOI: 10.1590/S1678-9946201860043.

[5] Brasil. Ministério da Saúde. Secretaria de Vigilância em Saúde. Coordenação-Geral de Desenvolvimento da Epidemiologia em Serviços. Guia de Vigilância em Saúde: volume único [recurso eletrônico]/Ministério da Saúde, Secretaria de Vigilância em Saúde, Coordenação-Geral de Desenvolvimento da Epidemiologia em Serviços. – 3ª. ed. – Brasília: Ministério da Saúde, 2019. 740 p. Available from: https://bvsms.saude.gov.br/bvs/publicacoes/guia_vigilancia_saude_3ed.pdf

[6] Pan American Health Organization/World Health Organization. Epidemiological Update: Arbovirus. Washington, D.C. PAHO/WHO. 2020. Available from: https://www.paho.org/en/documents/epidemiological-update-dengue-and-other-arboviruses-10-june-2020

[7] Mendonça FA, Souza AV, Dutra DA. Saúde pública, urbanização e dengue no Brasil. Sociedade & natureza. 2009:21(3):257-269. DOI: 10.1590/S1982-45132009000300003.

[8] Tauil PL. Aspectos críticos do controle do dengue no Brasil. Cadernos de Saúde Pública. 2002:18:3:867-871. DOI: 10.1590/S0102-311X2002000300030.

[9] Governo de Santa Catarina. Diretoria de Vigilância Epidemiológica, Vigilância e controle do *Aedes aegypti*. Orientações técnicas para pessoal de campo. Florianópolis – SC. 2019. Available from: http://www.dive.sc.gov.br/conteudos/publicacoes/Manual_completo_2019.pdf.

[10] Fares RCG, Souza KPR, Añez G, Rios M. Epidemiological scenario of Dengue in Brazil. BioMed Research International 2015(13). DOI: 10.1155/2015/321873.

[11] Quintanilha ACF. Caracterização clínica e epidemiológica de casos de dengue internados em hospital público de Campo Grande – MS. [thesis]. Federal University of Mato Grosso do Sul; 2010.

[12] Lima SFS, Moraes EC, Pereira G. Análise das relações entre as variáveis ambientais e a incidência de dengue no município de Campo Grande-MS. In: Simpósio Brasileiro De Sensoriamento Remoto – SBSR. 30 April to 5 May 2011. Curitiba: 2011. p. 28.

[13] Pacheco ICS, Carvalho AMA, Pontes ERJC, Silva, MGS. Relação entre

condições climáticas e incidência de dengue no município de Campo Grande, MS. Multitemas. 2017:22:51:235-252. DOI: 10.20435/multi.v22i51.1123

[14] Fundação Nacional de Saúde. Programa Nacional de Controle da Dengue (PNCD). Brasília: Funasa; 2002, 34p. Available from: http://bvsms.saude.gov.br/bvs/publicacoes/pncd_2002.pdf.

[15] Silva EMC, Cunha RV, Costa EA. National Dengue Control Program implementation evaluation in two border municipalities in Mato Grosso do Sul state. Epidemiologia e Serviços de Saúde. 2016:27. DOI: 10.5123/s1679-49742018000400007.

[16] Secretaria de Vigilância em Saúde. Dados e indicadores selecionados. Coordenação: Departamento de Análise de Situação de Saúde. Brasília: Ministério da Saúde; 2018.

[17] Siqueira Junior JB, Martelli CM, Coelho GM, Simplício ACR, Hatch DL. Dengue and Dengue Hemorrhagic Fever, Brazil, 1981-2002. Emerging Infectious Diseases 2005;11(1):48-53. DOI: 10.3201/eid1101.031091.

[18] Câmara FP, Theophilo RL, Santos, GT, Pereira, SRFG, Câmara DCP, Matos RRC. Estudo retrospectivo (histórico) da dengue no Brasil: características regionais e dinâmicas. Revista da Sociedade Brasileira de Medicina Tropical, 2007:40(2):192-196. DOI: 10.1590/S0037-86822007000200009.

[19] Guzmán MG, Kouri G. Dengue: an update. Lancet Infect Disease. 2002:2(1):33-42. DOI: 10.1016/s1473-3099(01)00171-2.

[20] Tsai TF, Vaughn D, Solomon T. Flaviviruses (Yellow Fever, Dengue Hemorrhagic fever, Japanese Encephalitis, St Louis Encephalitis, Tick-borne Encephalitis). In: Mandell G, Bennett J, Dolin R, eds. Principles and Practice of Infectious Diseases, 6th ed, Elsevier-Churchill-Livingstone, Philadelphia, 2005. DOI: 10.1590/S0036-46651992000200018.

[21] Machado V, Alessandra A, Negrão FJ, Júlio C, Estevan AO, Sales A, Brabes KCS. Custos diretos das hospitalizações por dengue: comparação entre o sistema de saúde público e privado em Dourados-MS. In: 49° MedTrop, 2013, Campo Grande. XLIX Congresso da Sociedade Brasileira de Medicina Tropical, 2013. p. 159.

[22] Valle D, Pimenta DN, Aguiar Z. Zika, Dengue e Chikungunya: desafios e questões. Epidemiologia Serviço de Saúde. 2016:25(2). DOI: 10.5123/S1679-49742016000200020.

[23] Neto ASL, Nascimento OJ, Sousa GS. Dengue, zika e chikungunya-desafios do controle vetorial frente à ocorrência das três arboviroses-parte I. Revista Brasileira Promoção em Saúde. 2016:29(3):305-312. DOI: 10.5020/18061230.2016.p305.

[24] Wilson ME, Chen LH. Dengue in the Americas. Dengue Bulletin. 2002:26:44-62. Available from: https://apps.who.int/iris/bitstream/handle/10665/163755/dbv26p44.pdf;sequence=1

[25] Swain S, Bhatt M, Biswal D, Pati S, Magalhães RJS. Risk factors for dengue outbreaks in Odisha, India: A case-control study. Journal of Infection and Public Health. 2020:13(4):625-631. DOI:10.1016/j.jiph.2019.08.015.

[26] Elaagip A, Alsedig K, Altahir O, Ageep T, Ahmed A, Siam HA, Samy AM, Mohamed W, Khalid F, Gumaa S, Mboera L, Sindato C, Elton L. Seroprevalence and associated risk factors of Dengue in Kassala state, eastern Sudan. Plos Neglected Tropical Disease. 2020:14(12). DOI: 10.1371/journal.pntd.0008918

[27] Ocampo CB, Mina NJ, Carabali M, Alexander N, Osorio L. Reduction in

dengue cases observed during mass control of Aedes in street catch basins in an endemic urban area in Colombia. Acta Tropica. 2014:132:15-22. DOI: 10.1016/j.actatropica.2013.12.019

[28] World Health Organization. Dengue and severe dengue [factsheet no. 117, revised January 2012]. Geneva, World Health Organization, 2012.

[29] World Health Organization. Guidelines on the quality, safety and efficacy of dengue tetravalent vaccines (live, attenuated). Geneva, World Health Organization. 2012.

[30] World Health Organization. Handbook for integrated vector management. Geneva, World Health Organization. 2012.

[31] World Health Organization. Global strategy for dengue prevention and control 2012-2020. [August 2012]. World Health Organization, 2012,

[32] Zara AL, Santos SM, Fernandes-Oliveira ES, Carvalho RG, Coelho GE. Aedes aegypti control strategies: a review. Epidemiologia e Serviços de Saúde. 2016: 25(2):391-404. DOI: 10.5123/S1679-49742016000200017.

[33] Erlanger TE, Keiser J, Utzinger J. Effect of dengue vector control interventions on entomological parameters in developing countries: a systematic review and meta-analysis. Medical and Veterinary Entomology. 2008:22:203-221. DOI: 10.1111/j.1365-2915.2008.00740.x.

[34] Souza TGB. Souza AP, Machado AM. Programa saúde na escola: Ações de combate ao mosquito *Aedes aegypti*. Editora Atena, Ponta Grossa- PR, 2019. DOI: 10.22533/at.ed.691203101.

[35] Chaebo G, Medeiros JJ. Is direct participation in co-production demogratic? Dengue vector control in Campo Grande – MS. Cadernos Gestão Publica E Cidadania, 2017:22:3-22. DOI: 10.12660/cgpc.v22n71.61067.

[36] Brasil, Ministério da Saúde. Secretaria de Vigilância em Saúde. Departamento de Vigilância Epidemiológica. Diretrizes nacionais para prevenção e controle de epidemias de dengue. Brasília: Ministério da Saúde; 2019. Available from: https://bvsms.saude.gov.br/bvs/publicacoes/diretrizes_nacionais_prevencao_controle_dengue.pdf.

[37] Brasil, Ministério da Saúde. Secretaria de Vigilância em Saúde. Departamento de Vigilância Epidemiológica. Guia de vigilância epidemiológica. 10. ed. Brasília: Ministério da Saúde; 2019.

[38] Barbosa JR, Barrado, JCS, Zara ASA, Siqueira Jr. JB. Avaliação da qualidade dos dados, valor preditivo positivo, oportunidade e representatividade do sistema de vigilância epidemiológica da dengue no Brasil, 2005 a 2009. Epidemiologia e Serviços de Saúde. 2015:24(1):49-58. DOI: 10.5123/S1679-49742015000100006.

[39] Callaway E. Mosquitoes engineered to pass down genes that would wipe out their species. Nature news 07 December 2015. DOI: 10.1038/nature.2015.18974.

[40] Santos KC, Siqueira Jr JB, Zara ALSA, Oliveira ESF. Avaliação dos atributos de aceitabilidade e estabilidade do sistema de vigilância da dengue no estado de Goiás, 2011. Epidemiologia e Serviços de Saúde. 2014: 23(2): 249-258. DOI: 10.5123/S1679-49742014000200006.

[41] Devine GJ, Perea EZ, Killeen GF, Stancil JD, Clark SJ, Morrison AC. Using adult mosquitoes to transfer insecticides to Aedes aegypti larval habitats. Proceedings of the National Academy of Sciences of the United States of

America. 2009:106(28): 11530-11534. DOI: 10.1073/pnas.0901369106.

[42] Brasil. Diagnóstico rápido nos municípios para vigilância entomológica do *Aedes aegypti* no Brasil (LIRAa): metodologia de avaliação dos índices de Breteau e predial. Brasília: Ministério da Saúde, Secretaria de Vigilância em Saúde, 2015. 60p.

[43] Sanchez L, Pelaez CO, Concepción D, Van der Stuyft P. Breteau index threshold levels indicating risk for dengue transmission in areas with low Aedes infestation. Tropical Medicine International Health. 2010:15(2):173-175. DOI: 10.1111/j.1365-3156.2009.02437.x.

[44] Shen JC, Luo L, Li L, Jing QL, Ou CQ, Yang ZC, Chen XG. The impacts of mosquito density and meterological factors on Dengue fever epidemics in Guangzhou, China, 2006-2014: a time-series analysis. Biomedical Environmental Science. 2015:28(5): 321-329.

[45] Caicedo-Torres W, Montes-Grajales D, Miranda WM, Agudelo MF. Kernal-Based machine learning models for the prediction of Dengue and Chikungunya morbidity in Colombia. Communications in Computer and Information Science. 2017:735(1). DOI: 10.1007/978-3-319-66562-7_34.

[46] Lagrotta MTF, Silva, WC, Souza-Santos R. Identification of key areas for Aedes aegypti control through geoprocessing in Nova Iguaçu, Rio de Janeiro State, Brazil. Cadernos de Saúde Pública. 2008:24(1):70-80, DOI: 10.1590/S0102-311X2008000100007.

[47] Feitosa FRS, Sobral IS, Silva MSF, Jesus EM. Estratégias de prevenção e controle da Dengue em Aracajú: Potencialidades e fragilidades. Caminhos de Geografia. 2016:17(60):149-168.

[48] Callaway E. The mosquito strategy that could eliminate dengue. Nature news 27 August 2020. DOI: 10.1038/d41586-020-02492-1.

[49] Hammond A, Galizi R, Kyrou K, Simoni A, Siniscalchi C, Katsanos D, Gribble M, Baker D, Marois E, Russell S, Burt A, Windbichler N, Crisanti A, Nolan T. A CRISPR-Cas9 gene drive system targeting female reproduction in the malaria mosquito vector Anopheles gambiae. Nature Biotechnology. 2016:34:78-83. DOI: 10.1038/nbt.3439.

[50] Gantz VM, Jasinskiene N, Tatarenkova O, Fazekas A, Macias VM, Bier E, James AA. Highly efficient Cas9-mediated gene drive for population modification of the malaria vector mosquito Anopheles stephensi. Nature Biotechnol. 2015:112(49):E6736-E6743. DOI: 10.1073/pnas.1521077112.

[51] Boyce R, Lenhart A, Kroeger A, Velayudhan R, Roberts B, Horstick O. *Bacillus thuringiensis* israelenses (Bti) for control of Dengue vectors: systematic literature review. Tropical Medicine International Health. 2013:18(5). DOI: 10.1111/tmi.12087.

Chapter 10

Environmental Manipulation: A Potential Tool for Mosquito Vector Control

Ukubuiwe Azubuike Christian, Olayemi Israel Kayode, Ukubuiwe Catherine Chinenye and Ugbede Bright Sule

Abstract

Mosquito borne diseases have continued to ravage man and his animals despite efforts to curb its spread. The use of chemicals has been the main thrust for control of all life stages of mosquitoes. Increased resistance to commonly used insecticides has called for renewed effort for vector control. Environmental management for vector control is one of the new strategies developed to tackle the menace of vectors. Manipulation of abiotic factors has widely gained acceptance due to laboratory and semi-field trials and findings. In this chapter, we reviewed literatures on some critical abiotic factors and their effects on bionomics and biological fitness of immature and adult life stages of mosquito species. We also looked at prospects for developing protocols based on these findings.

Keywords: *Aedes*, *Anopheles*, Biological Fitness, *Culex*, Vector Competence

1. Introduction

Mosquitoes are dipterans of the family Culicidae and are important in public health because of the bloodsucking habits of the females and transmission of important human diseases such as yellow fever, malaria, filariasis, and dengue [1]. Mosquitoes have three subfamilies namely, Anophelinae, Culicinae, and Toxorhynchintae. Among these, only Anophelinae and Culicinae contain medically important Genera (*Aedes, Anopheles,* and *Culex*) that are efficient in disease transmission [2]. The discovery of the role played by mosquitoes in disease transmission and the need to develop cost-effective and species-specific control measures, through sound understanding of the biology and ecology of these vectors, have stimulated interest in mosquito research [3, 4].

Mosquitoes have four life stages (egg, larva, pupa, and adult). The egg, larval (comprising of four instars), and pupal stages are all aquatic. Collectively, these stages take about 7 to 14 days to complete development, depending on ambient temperature, as they are cold-blooded. Apart from temperature, other factors also affect the developmental times of mosquitoes. These include photoperiod, density, feed quality and quantity, salinity, hardness, nitrate and sulphate contents, and water pH.

The behaviour of mosquitoes determines their importance/ status as nuisance insects or pathogen vectors, therefore, governs the selection of control methods [5, 6]. Most female mosquitoes depend on blood from animals or humans for maturation of their eggs [7]. Species that prefer to feed on animals are usually not very effective in transmitting human diseases [8–10], while those that rest indoors are usually the easiest to control [11].

Mosquitoes are the most important insect of public health concern, basically due to their nuisance, ferocious and infective female bites. Diseases spread by female bites, for example, malaria have been responsible for millions of deaths, especially, of pregnant mothers, children below the age of five, and immune-compromised individuals [12, 13]. The brunt of the scourge of mosquito vector-borne diseases (MVDs) is exceptionally high in Tropical and Subtropical climates. This is due in part to clemency of the weather conditions, rapid urbanisation, high anthropogenic activities, proliferation of suitable breeding habitats, among others [14, 15]. In Temperate climes, diseases such as malaria has been eradicated, although, there are risks of reintroduction due to increased human movement and climate change [16]. However, some MVDs, especially, those transmitted by Culicines (e.g., West Nile fever, Dengue, and Chikungunya) are still prevalent and are of significant public health concern in these countries [17]. There is, therefore, need for effective, efficient and eco-friendly vector control tool for mosquito eradication and prevention of disease.

Despite concerted efforts towards mosquito eradication, multi-faceted epidemiological factors have impeded the complete eradication of MVDs, especially in low income countries. These factors include, but are not limited to lack of social and political will from stakeholders, poor budgetary allocations, insufficient manpower, variation in the biology of principal and secondary vectors of the disease. Among these factors, the biology of vector species has received numerous attentions with a lot of scientific publications on the subject matter. In fact, sound knowledge of the biology of mosquito vectors is important for successful implementation of control intervention protocols. This is, exceptionally so, as spatio-temporal variations in biology and genetics of species complex and sibling species [18].

The use of chemical insecticides has been man's foremost weapon against these vectors. However, increased incidence of (cross- and class-) resistance to insecticides by most vector species, with the attendant environmental concerns have necessitated significant reduction in the application or overall ban of some chemical insecticide formulations [19]. There is, therefore, need to develop an alternative robust protocol that would be devoid of the pitfalls of chemical insecticides. Among these alternatives is the integrated vector management (IVM), based on Environmental Management of Vectors (EMVs). The EMVs has proven to be most effective and efficient in this regards [20].

2. Chemical control, the Main thrust of mosquito vector control

Chemical control of mosquito vectors involves the use of chemicals, either synthetic or organic to reduce vector population or contacts with human hosts or their animals. Chemical agents that reduce mosquito vector populations are referred to as chemical insecticides and are usually designed to target various life stages of the mosquitoes. Hence, there are ovicides, larvicides, pupicides, and adulticides, for the control of eggs, larvae, pupae and adult life stages, respectively. Other chemicals produced for control include repellents, oviposition deterrents, among others. Chemical insecticides used in various forms such as aerosols, indoor residual sprays, impregnated household materials, repellents, larvicides among others have contributed, substantially, to the reduction of MVDs in most countries. The introduction and improvement of these chemical agents have mitigated the disease burdens by reducing the vector population.

Despite the initial gains of chemical insecticides in global eradication of vector-borne diseases, cases of increased insecticide resistance have been reported globally. Mosquitoes have developed class- and/or cross-resistance to insecticides, with various mechanisms of resistance (metabolic, physiological, anatomical, or behavioural). Further, environmental studies have shown persistence of some of these chemicals in the environment over decades [21], with cases of destruction and even elimination of non-target and beneficial organisms in the ecosystem. Some degree of mammalian toxicity has also been reported for some of these insecticides. These pitfalls of chemical control methods have necessitated the call for the development of other environmentally safe control protocols which will be efficient and effective in reducing vector population and, hence vectored diseases.

Several alternatives to chemical control methods have been developed against mosquitoes. These include biological, cultural, legal, and genetic control methods. However, the need to integrate these diverse control strategies towards reducing vector population has given rise to Integrated Mosquito Management (IMM).

3. Environmental Management in Integrated Vector Control

A successful larval control protocol requires adequate knowledge of breeding ecology of the vector including, the developmental environmental requirements of the immatures [22–24].

The World Health Organisation (WHO) in 1982, developed the EMVs - a subset of the Concept of Integrated Vector Control - as a roadmap for the control of major vectors and intermediate hosts of diseases. Environmental management activities for vector control involves planning, organisation, carrying out and monitoring activities for the modification and/or manipulation of environmental factors or their interaction with man to prevent or minimise vector propagation and reducing man-vector-pathogen contact [25–27]. Environmental Management of Vector (EVMs) involves three strategies, namely, environmental modification, environmental manipulation, and modification and/or manipulation of human habitation or behaviour to reduce vector-human contact [20].

Environmental Modification involves physical alterations of land, water and vegetation, which are usually permanent or long–lasting, aimed at preventing, removing or reducing the habitats of vectors without causing unduly adverse effects on the quality of the human environment. Activities enlisted under this include drainage, filling, land levelling and transformation and impoundment margins [20]. The second aspect, Environmental Manipulation, consists of any planned recurrent activity aimed at producing temporary conditions unfavourable to the breeding of vectors in their habitats (this is the focus of this chapter). Strategies involved include water salinity changes, stream flushing, water level regulation in reservoirs, dewatering or flooding of swamps or boggy areas, vegetation removal, shading and exposure to sunlight [25].

While, the third aspect, modification of human habitation and/or behaviour, involves strategies that reduce man – vector–pathogen contact. Examples of this approach include siting of human settlements away from vector sources, mosquito proofing of houses, personal protection and hygiene measures against vectors. Others include provision of mechanical barriers and facilities for water supply, wastewater and excreta disposal, laundry, bathing and recreation to prevent or discourage human contact with infested waters, and zoo-prophylaxis, the placement and provision of an alternate blood meal source to divert vectors away from the human blood source [26].

Environmental modification and human habitation and/or behaviour modification have been fully investigated and the outcomes implemented. These results have resulted in major reductions in the epidemiology of the diseases transmitted by some vectors, generally, and mosquitoes in particular. Yet, the diseases vectored by mosquitoes continue to ravage mankind due to either changes in vectors' biology over time, or ineffectiveness of these methods to fit into current trends of Integrated Mosquito Management (IMM) protocols.

Environmental manipulation techniques, on the other hand, provide a sustainable remedy to mosquito vector control, as its ultimate goal is to produce adult mosquitoes which are less fit as vectors by changing the quality of established mosquito breeding habitats. However, environmental manipulation approaches have not been fully exploited, especially, in terms of changing vital developmental components/factors of the vector's environment to reduce biological fitness. Such strategies are promising as they are always targeted at the weakest link (larvae) in development of the vector.

Although these developmental factors act together (antagonistically or synergistically) to affect the growth of mosquitoes in the wild, laboratory studies have shown their individual contribution to vector success. It is hoped that manipulating these developmental components in the wild will produce adult life stages that are less fit as vectors, hence, disrupting the chain of disease transmission [26, 28].

4. Environmental manipulation of mosquito habitats

Integrated Mosquito Management (IMM) protocols based on manipulation of vector's micro-habitat, especially during development, is promising to be an effective strategy in the control of the major disease-causing vectors. The goal of this control approach (Environmental Manipulation) is not to eradicate mosquitoes from the surface of the earth, as it is often advocated, but to identify the environmental factor (s)/variable (s) that contribute (s) to their success, manipulate them to the extent of producing mosquito species which are not fit as vectors of diseases.

Understanding species-specific effects of environmental factor on the bionomics of mosquitoes will be valuable in developing control protocols [20, 26]. One advantage of such protocol is that it will target the weakest link (larvae) during development. Apart from higher vulnerability to toxic materials, larvae are confined within the water body and do not migrate away from toxic environment, unlike adult life stage. More so, application of such protocol would either be species-specific or broad-ranged, depending on the specific developmental requirements of the vectors, and would not require special expertise or training.

5. Abiotic components of mosquito habitats

In mosquito ecology, abiotic components are sometimes referred to as physio-chemical factors, and include water conditions such as pH, salinity, hardness, alkalinity, temperature, sulphate, nitrate and phosphate contents, etc. These factors affect mosquito larval bionomics in diverse ways and determine spatio-temporal abundance, distribution and biological fitness of mosquito species. Extensive studies have been carried out either in combination or as a single factor on influence of some of these factors on mosquito biology. With some studies transcending larval bionomics to adult bionomics and filial generations. Mosquitoes have shown limits of tolerance, with some degree of adaptability to these factors.

In nature, physio-chemical factors interact in diverse ways to affect the phenotype and genotype of mosquito species, there is, therefore, need for semi-field and field trials of results from laboratory studies. Further, since there is 'no physiology of mosquito' [29], influence of abiotic factors on specific vector's bionomics is key in understanding their roles in mosquito development and disease transmission.

6. Influence of selected abiotic factors on bionomics of mosquitoes

For the sake of this chapter, a concise and systematic review of the contributions of some abiotic factors to the development of mosquitoes and the possibility of developing novel control intervention will be taken. More elaborate discussions of the subject matter can be found in other publications. Further, the review will be on critical indices of disease transmission in mosquito: duration of development, larval growth rate, immature survivorship, number of adults at emergence, adult longevity and survivorship, wing-based indices, body size, and metabolic reserves.

Duration of development is the time taken to complete pre-imaginal life stages (i.e., from egg to pupae). This is critical to biological fitness as longer developmental times affect resource mobilisation and reduce vector-host contact frequencies [30], among others. Larval growth rate indicates the daily rate of biomass accumulation during the photoperiod (larval stage). It estimates daily weight gain as larvae which is critical to successful adult life traits [31]. Immature survivorship is an index of developmental success of a species, and indicates maternal reproductive success and ensures generation continuity [32].

Number of adults at emergence and sex ratio determines mating frequencies, time before sexual maturity (in male mosquitoes). These are critical for swarming, host-seeking and laying of fertile eggs. Adult longevity and survivorship is an indication of life span of mosquito when fed either energy source alone (sucrose) or in combination with blood meal [32]. These are crucial for extrinsic incubation period within female mosquitoes; disease pathogens complete development in longer-lived adult vector.

Wing-based indices include measurements like wing length, surface area and fluctuating asymmetry. In mosquito physiology, wing length is a proxy for entomological indices such as body size, weight, fecundity [33, 34], longevity [35], host-finding success [36], blood-feeding success [37], survivorship [38] and vectorial capacity; all these influence biological fitness of the vector for disease parasite transmission. Fluctuating asymmetry (FA) is commonly used as a measure of stress during development, and fitness of adult mosquitoes [39, 40].

The body size of adult females influences the number of blood meals acquired and required to complete the first gonotrophic cycle and the number of eggs [41] as smaller females take longer to achieve reproduction and produce fewer off-springs. Metabolic reserves of epidemiological interests are protein, lipid, glucose and glycogen. Most female mosquitoes require blood meals to provision and mature eggs (i.e. fecundity), however, the first ovarian cycle is determined by metabolic reserves, especially that derived from larval nourishment [42]. Autogenous mosquitos do not require blood meals to lay the first few batches of eggs, unlike anautogenous species, where blood meal in addition to larval-derived nutrient reserves is a prerequisite to laying eggs [43, 44]. Metabolic reserves of newly emerged adults (teneral reserves) affect important female reproductive processes, such as the utilisation of reserves, fecundity, longevity, flight, the formation of new tissues and organs, and blood meal consumption and utilisation [45].

6.1 Water temperature

As cold-blooded organisms, mosquitoes rely on environmental temperature for all metabolic life processes. Literature on influence of temperature on mosquitoes abounds, however, an attempt will be made to summarise these for the sake of this chapter. Temperature zones of the world have been broadly categorised into Tropical/Sub-tropical and Temperate climates. Tropical and sub-tropical climates have relatively high temperatures, while temperate climates have colder to freezing temperatures. The response of mosquitoes from these climes to temperature are different, hence, predictions on development and biology based on prevailing temperatures would be also different [46, 47]. In the tropics and sub-tropics, with all-year-round favourable developmental temperatures, the influence of temperature is extremely strong. Apart from facilitating all year proliferation of mosquitoes, these temperatures also favour development of parasites within the vectors. In temperate climes, however, where extremely cold to freezing temperatures abound, mosquito development is often slow, and at times, halted during adverse weather conditions. These also affect the development of pathogen in the vectors [17].

Information on the influence of temperature on fitness indices at both immature and adult life stages can be employed in developing novel control strategies. Temperature change during development in mosquitoes produce different phenotypes [48–52], endowed with different genotypes. Future genetic studies can be based on these phenotypes.

Mosquitoes are adapted to surviving and reproducing at specific temperature ranges and, thus, have different thermal limits. Temperatures outside these limits lead to disruption of biological processes, or often death. Exposure to high temperatures result in denaturing of proteins, alteration of cell membranes, enzyme structures and properties, pH and ion concentrations, destroying wax complex of the cuticle, leading and desiccation due to evaporation [53].

Genera and species differential responses to lethal temperatures at given time ranges have been reported. Ambient temperature affects mosquito proliferation [54, 55]. Colder temperatures delay embryo eclosion, reduce hatch rate [56], and developmental rate [57]. While higher temperatures elicit faster pupation [58–60], reduced ecdysis [61] and longevity [62].

Olayemi *et al.* [48] reported shorter duration of development in *Cx. quinquefasciatus* mosquito, with increase in temperature: taking as low as 6 days at 34°C. This was, however, accompanied by reduced immature survivorship at temperatures above 30°C for the species. Ukubuiwe *et al.* [49] reported a linear relationship between temperature increase and growth rates and duration of development in *Cx. quinquefasciatus* and a negative relationship between temperature and immature life stages' survivorship. Similar observations have been recorded for *An. gambiae* [57], *Ae. krombeini* [63], and *Cx. tarsalis* [64].

Although high temperatures are associated with faster development, several studies have observed significant reduction in the number of adults at emergence and longevity (with increasing temperature) in mosquitoes such as *Cx. tarsalis* [65], *An. gambiae* [62], *An. sergentii* [66], *Ae. albopictus* [61] and *Cx. quinquefasciatus* [49]. Altering developmental temperature above the upper thermal limit will, therefore, increase immature life stage mortality, reduced adult emergence and human-vector contacts. Temperature change also affect post-emergence longevity in mosquitoes. Adult *Cx. quinquefasciatus* mosquitoes lived the longest at 30°C, whereas, at 34°C, longevity was significantly reduced [49]. High temperatures reduce adult daily survivorship in *Cx. apicinus* and *Cx. hepperi* [67]. In *Cx. quinquefasciatus,* female mosquitoes survived more than the male species at all temperatures investigated [49].

Wing lengths reduce with increase in temperature. This is, however, species-specific. In *Cx. quinquefasciatus,* temperatures above 30°C reduced ptero-fitness [48]. Other researchers have reported similar temperature-dependent variation in adult wing lengths in *An. merus* [68], *An. quadrimaculatus* [50], *Cx. apicinus* [67] and *An. dirus* and *An. sawadwongporni* [51].

Body parts of immature mosquitoes are also affected by water temperature change. Higher water temperatures reduced larval body length, width, and volume and pupal cephalothoracic length in *Cx. pipiens* mosquitoes [40], *An. merus* [68], and *Cx. quinquefasciatus* [52].

There are scanty published data on the influence of temperature on metabolic reserves in mosquitoes. In *Cx. quinquefasciatus,* mosquitoes reared at 34°C had the lowest of all metabolic reserves, while higher reserves were recorded at 30°C [49]. Therefore, for this species, biological fitness of this species is enhanced at 30°C. Therefore, techniques designed to increase developmental temperature above this, will significantly reduce fitness of this vector.

6.2 Water pH

Another important immature breeding factor is water pH. Level of water pH level depends on its carbonic acid equilibrium. Just as for temperature, there are permissible tolerance limits for most aquatic organism, including mosquitoes. Outside these limits, developmental processes and normal physiology are affected [69–71]. Water pH is affects availability of essential mineral and food elements for development of mosquitoes, thereby, distribution [72], and survivorship of mosquitoes [73]. Field studies have reported strong positive correlation between pH and the quality of mosquito habitats [72, 74]. Studies have also shown that mosquito larvae adapt to and tolerate fluctuations in ionic levels in these habitats [75–77].

Even though water pH regulates growth in mosquito species, adaptation has been reported in some vector species. For Aedine mosquitoes, species-specific tolerance ranges have been reported [74, 78]. *Culex pipiens* showed limited survivorship at pH 4.4 to 8.5 [79]. At extreme pH values of 4.0 and 10.0, *Cx. quinquefasciatus* had reduced developmental successes and adult biological fitness indices [28]. Its optimum range for development is between pH 5.0 and 8.0. From these studies, it seems that not all habitats in the wild supports development of mosquitoes, however, those that do, might actually reduce biological fitness.

In *Ae. aegypti,* percentage emergence reduced at pH 3.6 and 4.2 [78]. In *Cx. quinquefasciatus,* however, immature survivorship was highest between pH 5.0 and 8.0 and lowest at pH 10.0 [28]. According to the authors, male mosquitoes were most affected by the change in pH levels, and adult survivorship was, exceptionally, high between pH 5.0 to 8.0 and lowest at pH 4.0. Laboratory investigations have revealed larval-age-related increase in reserve accumulation in *Cx. quinquefasciatus* at different water pH level. Rate of mobilisation and accumulation of metabolic reserve components were reduced at extreme water pH conditions and highest at pH 7.0 [28].

6.3 Water hardness

Water hardness levels also play epidemiological roles in regulating the occurrence and distribution of mosquito species. It also determines the quality of mosquito larval habitat [80]. Mostly, the hardness of water is determined by the nature of the topsoil and the presence or absence of divalent cations of calcium (Ca^{2+}), magnesium (Mg^{2+}), ferrous iron (Fe^{2+}) and manganese (Mn^{2+}) ions [81]. Ample evidence has been shown that these ions play protective, metabolic, structural, and physiologic roles in aquatic organisms [82, 83]. Water hardness levels have been categorised into 'soft', 'slightly hard', 'moderately hard', 'hard' and 'very hard' water with calcium trioxocarbonate ($CaCO_3$) content, respectively, less than 17.0, 17.0–59.0, 60.0–119.0, 120.0–180.0, and greater than 180 mg/L [84].

Despite the importance of water-hardness-causing ions, mosquitoes perform optimally within set limits [40]. Outside these limits, their occurrence, distribution, physiology, growth and development will be greatly affected. Calcium ions, for example, in excess elicit environmental stress conditions and affect feeding rates of aquatic organisms [85]. Impaired feed intake affects the amount of energy readily available for normal activities of the organisms, and mobilisation from one stage of life to the other [86, 87].

The effects of water hardness on mosquito bionomics have mostly been extrapolated from field data. Field data have suggested that 'moderately' and 'hard' water support mosquito growth [88–93]. Species-specific data on the influence of water hardness is important in elucidating actual contributions to mosquito growth. Hence, Ukubuiwe *et al.* [28, 94] and Aminuwa *et al.* [95] reported the influence of water hardness level on development and morphometric of *Cx. quinquefasciatus*. These authors reported reduced immature developmental successes and adult biological fitness indices of the species at hardness levels ≥150 mg/L CaCO3 (i.e., above 'hard' water). Duration of development of the species was fastest in 'soft' water, but longest in 'very hard' water [28, 95]. Larval growth rates were also highest in 'moderately hard' water but lowest in 'very hard' water [28]. This suggests growth-regulating effect of water hardness on the species, especially when in high quantities.

Further, first instar larvae of *Cx. quinquefasciatus* were most affected by water hardness level change, while immature survivorship of the species was lowest in 'very hard' water. Adult survivorship of *Cx. quinquefasciatus* were highest in 'moderately hard' water $CaCO_3$ [28]. This study explained the relatively poor productive of habitats with high water 'hardness' content [80] and absence of species in some habitats with 'very hard water' (personal field observations). 'Very hard' water conditions produced very small-sized *Cx. quinquefasciatus* mosquitoes (across all life stages). Wing-based fitness indices for the species were lowest in 'very hard' water [94]). The author concluded that moderately hard water conditions are the best for overall fitness and performance of the species.

Information on the influence of water hardness conditions on metabolic reserves of mosquitoes is also scanty. However, laboratory investigations suggest that as water hardness level increased, mobilisation and utilisation of reserve components increased, resulting in depletion of adult life stage values [28]. These findings have epidemiological implications on population density and degree of human-mosquito contacts for disease pathogens transmission. Therefore, protocols involved in changing hardness levels of mosquito habitats will produce less biologically fit mosquitoes, thereby reducing scourge of the diseases transmitted by these vectors.

6.4 Light duration (photoperiod)

Photoperiod regulates most physiological processes in insects, including growth, diapause and longevity [96–98]. Many insects respond differently to length of day (photoperiod) and depend on it as cue to seasonal development [99–103]. Some insect species develop faster under short day-length, while in others, development is almost halted, and for a few, the insects were indifferent to variation in the day-length [104]. Different mosquito phenotypes are produced on exposure to different photoperiod regimens. Such information could be harnessed for producing small-sized, biologically less fit mosquitoes.

An avalanche of literature exists on the influence of photoperiod on bionomic of mosquitoes. In *Cx. pipiens,* short photoperiodic conditions cause the development of smaller ovarian follicles [105]. These also increase the lifespan in *An. quadrimaculatus* [106], wing length, areas, body weight and greater wing area per unit body weight in *Cx. pipiens pipiens* [107]. In *Toxorhynchites rutilus,* long day-lengths promotes rapid growth and metamorphosis, while short days retards development [108]. In *Wyeomyia smithii* [99], *An. quadrimaculatus* [106, 109], and *An. crucians* [110], short day-lengths has been associated with higher survivorship.

In an extensive study on influence of photoperiod on indices of fitness in *Cx. quinquefasciatus,* Ukubuiwe *et al.* [111] reported shorter durations of development, and higher numbers of adult emergence at short light durations. Exposure of larvae of *Cx. quinquefasciatus* to longer photophase produced phenotypes with shorter wing length and higher fluctuating asymmetry [111]; representing possible developmental stress. Metabolic reserves of the species were also affected by photoperiod conditions; larvae reared at short photoperiods had the highest biomass accumulation. Longer light duration reduced life stages' metabolic reserves and their caloric indices. Larvae of the species reared at longer photo-phase required relatively more metabolic components for pupation and pupal eclosion than those reared in shorter day lengths [112]. Based on the above-mentioned positive influence of short day-lengths on larval and adult fitness indices, it can be concluded that mosquitoes from shorter photophase may prove to be better vectors than those from longer photophase. As these spent the shortest time for development, survived most, were bigger, and accumulated more metabolic reserve than those from longer photoperiodic conditions.

The knowledge and information generated from studies on the effects of photoperiod on vector biology can be incorporated in control strategies that may either retard or slow down the developmental processes or produce less fit adults. Further laboratory studies on vector competence of mosquito species exposed to various photoperiodic regimens are also advocated.

6.5 Larval density/overcrowding

Larval density, described as the number of mosquito per unit, has profound effects on the life cycle of mosquitoes. Most field surveillance of vector species incorporates larval density, often expressed as number of larvae per dip, during larval sampling [113]. Measuring larval density in this regards is generally employed during pre-and post-intervention procedures. The data generated from these exercise tell little or nothing about the contribution of overcrowding to biological fitness of mosquitoes. Through laboratory studies, however, remarkable influence of larval density on various immature and adult life attributes have been elucidiated. Such information though laboratory-based, reveals the contribution of larval density and its possible inclusion in developing novel control strategies. Such information will also assist in making informed decisions and the deployment of scarce resources in vector control programs.

High larval density has been associated with various degrees of competitions [114, 115], resulting in phenotypic plasticity in *An. arabiensis* and *An. gambiae* [116] and *Cx. quinquefasciatus* [117]. High larval densities also affect the following indices of biological fitness; immature and adult survivorship, population quality, eclosion rates [118], sex ratio [119], and vector competence [120].

High larval density negatively affects the size and quality of adult *An. gambiae* mosquitoes [121]. According to Ye-Ebiyo *et al.* [122], overcrowded larvae of *An. arabiensis* are often smaller and short-lived as adults. More so, increased larval density has also been linked to sex-specific reactions such as parasite infection [123] and larval mortality in *An. stephensi* [124] and *Ae. aegypti* [125]. Overcrowding conditions increase larval development times [121], but reduce the size of the mature larva, pupa and resulting adult [126], with its attendant effects on the fecundity of females [127].

In *Cx. quinquefasciatus,* fourth larval instars were most affected by increasing larval density, and resulted in reduced rate of larval growth and immature life survivorship of the species [126]. These authors opined that high larval densities induce stress and reduced feed intake due to competition, with reduction in metabolic reserves. A similar reduction in growth rates have been reported in *Ae. albifasciatus* [128] and *Ae. aegypti* [129].

Although, in nature, gravid mosquitoes tend to avoid ovipositing in habitats that will pose serious developmental tasks on the young larvae as is seen in *An. gambiae* [121]; these laboratory studies explain what might happen when several mosquito species oviposit in habitats, with initial favourable growth factors, which gets exhausted over time.

High larval density significantly affects emergence success, adult survivorship, and longevity in most mosquito vectors: *An. stephensi* [124], *Ae. sierrensis* [130], *Ae. albopictus* [131] and *Ae. aegypti* [132], *Cx. quinquefasciatus* [126, 133], *Cx. pipiens fatigans* [134, 135], *Cx. tarsalis* [136], and *Wyeomyia smithii* [137]. High density also reduce female fecundity in *Ae. aegypti* [125, 138] and *An. gambiae* s.s [131, 139]. Similarly, negative effects of high density have also been reported in *Ae. aegypti, Cx. pipiens, An. albimanus*, and *An. gambiae* [118], and *Cx. sitiens* [140]. Further, during metamorphosis *Cx. quinquefasciatus* larvae in crowded environments expended more energy for pupation and eclosion. The reason for this is not clear but will lead to depleted energy reserves for adult's life attributes [126].

These observations may imply that higher mosquito larval densities of may not necessarily suggest potential health threat as often reported; as such mosquitoes would have undergone developmental stress, which affects post-immature life traits. Based on the above-mentioned studies, these mosquitoes manifested evidence of developmental stress, such as high fluctuating asymmetry, hence, may be 'bad' fliers, unable to secure mate and forage. More so, these mosquitoes may not live long enough to transmit pathogens, and may not have adequate energetic budgets for intra-vectoral pathogen development as greater energy reserves have been expended for metamorphosis, coupled with low adult survivorship and reduced longevity. Though there are no documented evidence on these submissions, further studies are advocated. However, if the above scenario permits in the wild, it may be nature's way of regulating mosquito population explosion, among others.

7. Prospects

Despite the laboratory evidence from the study of abiotic factors on developmental and adult fitness indices, further studies, either semi-field or field experiments to concretise these observations to enable full integration into control protocol. More so, for effective incorporation of these protocols, the following are gaps of knowledge to be filled.

- **Genetic bases of vector-abiotic factors interaction**

  Phenotypic expressions usually have genetic undertones. Genetic studies to decipher the genetic bases of the phenotypes observed in the studies above are highly recommended. The information generated will be useful in genetic manipulations to produce less fit (selective disadvantaged) mosquitoes.

- **Other Abiotic factor**

  The influence of other abiotic factors such as sulphate, nitrate, alkalinity, Dissolved oxygen, should also be investigated.

- **Vector competence**

  The studies highlighted above did not elucidate the influence of the abiotic factors on the ability of the mosquitoes to develop and transmit disease pathogen. Even though vector competence in mosquitoes has been inferred from the results, further investigated on this is key is understanding the aspect of vectors' physiology.

- **Reproductive performance**

  There is also a need to conduct scientific studies on the reproductive performance of females from the regimens of the factors investigated.

- **Generational effects**

  Most of the studies reviewed in this chapter ended with the parent stock. No scientific report exists for the influence on the progenies. It would be meaningful to study the effects of the factors on the development and adult fitness indices of subsequent generations. This will provide information on the sustainability of the protocols when developed.

- **Developing Predictive Modelling**

  Predictive models will assist in developing protocols to forecast influence of prevailing environmental factors on the biological fitness of mosquitoes, especially in high risks area for maximising control costs.

8. Conclusion

Single vector control approach such as insecticide application has failed to curb the spread of mosquito borne disease and has necessitated development of other techniques such environmental manipulation. Focusing on changing the quality of mosquito breeding habitats to produce less biologically fit adults, in capable of transmitting disease pathogen, environmental manipulation is promising to curb vector population explosion and disease transmission.

9. Recommendation

For effective implementation of environmental manipulation for mosquito vector control, field and semi-field trials of the influence of these critical abiotic factors should be considered. Genetic studies on bases of vector-abiotic factors interactions should be investigated. Other abiotic factors, other than those mentioned in this chapter should be studied. Influence of various abiotic factors on mosquito vector competence and reproductive performance should be explored. The effects of these critical abiotic factors on future generations of parent stocks should also be investigated. Finally, with the information generated, predictive models can be developed.

Acknowledgements

We wish to thank Miss Favour Ngoziala for typing the drafts of the manuscript. Most of the studies reported by the Author was sponsored by a grant from United State Agency for International Development (USAID), Higher Education Partnership/University of Mississippi (HEP/UM) (UM SUB-AWARD NO. 15-12-024).

Conflict of interests

The authors declare no conflict of interests.

Author details

Ukubuiwe Azubuike Christian[1*], Olayemi Israel Kayode[1], Ukubuiwe Catherine Chinenye[2] and Ugbede Bright Sule[1]

1 Applied Entomology Unit, Department of Animal Biology, Federal University of Technology, Minna, Nigeria

2 Department of Microbiology, Federal University of Technology, Minna, Nigeria

*Address all correspondence to: a.ukubuiwe@futminna.edu.ng

References

[1] El-Badry AA, Al-Ali KH. Prevalence and seasonal distribution of dengue mosquito, *Aedes aegypti* (Diptera: Culicidae) in Al-Madinah Al-Munawwarah, Saudi Arabia. Journal of Entomology. 2010;7 (2):80-88.

[2] Paulraj MG, Reegan AD, Ignacimuthu S. Toxicity of Benzaldehyde and Propionic Acid against Immature and Adult Stages of *Aedes aegypti* (Linn.) and *Culex quinquefasciatus* (Say) (Diptera: Culicidae). Journal of Entomology. 2011;8:539-547.

[3] Gyapong M, Gyapong JO, Adjei S, Vlassof C, Weiss M. Filariasis in Northern Ghana: Some cultural beliefs and practices and their implications for disease control. Social Science and Medicine. 1996;43:235-242.

[4] Ahorlu CK, Dunyo SK, Koram, KA, Nkurmah FK, Aagaard-Hansen J, Simonsen PE. Lymphatic filariasis related perceptions and practices on the coast of Ghana: implications for prevention and control. Acta Tropica. 1999;73:251-264.

[5] Hill SR, Hansson BS, Ignell R. Characterization of Antennal Trichoid Sensilla from Female Southern House Mosquito, *Culex quinquefasciatus* Say. Chemical Senses. 2009;34(3):231-252.

[6] Syed Z, Leal WS. Acute olfactory response of *Culex* mosquitoes to a human- and bird-derived attractant. Proceedings of the National Academy of Sciences. 2009;106(44):46-50.

[7] Tadel WP, Thatcher BD. Malaria vectors in the Brazilian Amazon: *Anopheles* of the subgenus *Nyssorhynchus*. Rev. Institute of Medicine of the Tropics. 2000; 42 (2):287-300.

[8] Elissa AH, Nicole FA, Zwiebel LJ, Carlson JR. Olfaction: Mosquito receptor for human-sweat odorant. Nature. 2004;427(6971):212-213.

[9] Davis JK. *UC Davis Researchers Identify Dominant Chemical That Attracts Mosquitoes to Humans*. University of California. 2009; Accessed 2011-07-06.

[10] Devlin H. Sweat and blood why mosquitoes pick and choose between humans. *London: The Times*. 2010; Acessed May 13, 2010.

[11] Olayemi IK, Ande AT, Isah B, Idris AR. Epidemiology of malaria in relation to climatic variables in Minna, Nigeria. African Journal of Medical Sciences. 2009; 2 (1): 5-10.

[12] World Health Organization, WHO. World Malaria Report: 2020. Geneva: World Health Organization; 2020a.

[13] World Health Organization, WHO Global programme to eliminate lymphatic filariasis: progress report, 2019. Geneva: World Health Organization; 2020b

[14] Nwoke BEB, Dozie INS, Jiya J, Saka Y, Ogidi JA, Nuttal I. The prevalence of hydrocele in Nigeria and its implication on mapping of lymphatic filariasis. Nigerian Journal of Parasitology. 2006;27:29-35.

[15] Terranella A, Eigege A, Jinadu MY, Miri E, Richards FO. Urban lymphatic filariasis in central Nigeria. Annals of Tropical Medicine and Parasitology. 2006;100(1):1-10.

[16] Centers for Disease Control and Prevention, CDC. Malaria: Malaria Transmission in the United States. 2020. Centers for Disease Control and Prevention. 2020.

[17] European Centre for Disease Prevention and Control, ECDC. Climate

change in Europe: Vector-borne disease. European Centre for Disease Prevention and Control. 2020.

[18] Touré YT, Petrarca V, Traoré SF, Coulibaly A, Maiga HM, Coluzzi M. The distribution and inversion polymorphism of chromosomally recognized taxa of the *Anopheles gambiae* complex in Mali, West Africa. Parassitologia, 1998;40:477-511.

[19] Knox TB, Juma EO, Ochomo EO, Pates Jamet H, Ndungo L, Chege P, Bayoh NM,N'Guessan R, Christian RN, Hunt RH, Coetzee M. An online tool for mapping insecticide resistance in major Anopheles vectors of human malaria parasites and review of resistance status for the Afrotropical region. Parasit Vectors. 2014;7:76.

[20] World Health Organisation, WHO. Manual on Environmental Management for Mosquito Control with special emphasis in Malaria Vectors. 1982;66:140-148.

[21] Gill HK, Garg, H. Pesticides: Environmental Impacts and Management Strategies, Pesticides – Toxic Aspects, Marcelo L. Larramendy and Sonia Soleneski, 2014; IntechOpen, DOI: 10.5772/57399.

[22] Singh RK, Dhiman RC, Singh SP. Laboratory studies on the predatory potential of dragon-fly nymphs on mosquito larvae. Journal of Communicable Diseases. 2003;35(2): 96-101.

[23] Marten GG, Reid JW. Cyclopoid copepods. Journal of the American Mosquito Control Association. (2007);23(2); 65-92.

[24] Olayemi IK, Ande AT. Relative abundance and seasonal distribution of adult mosquito species in Offa, Kwara State, Nigeria. Nigerian Journal of Entomology. 2008; 25:47-52.

[25] Ault SK. Environmental management: a re-emerging vector control strategy. American Journal of Tropical Medicine and Hygiene. 1994; 50:35-49.

[26] Bond JG, Rojas JC, Arredondo-Jime´nez JI, Quiroz- Martı´nez H, Valle J, Williams T. Population control of the malaria vector *Anopheles pseudopunctipennis* by habitat manipulation. Proceedings of the Royal Society London Series B, Biological Sciences. 2004;271: 2161-2169.

[27] Randell HF, Dickinson KL, Shayo EH, Mboera LEG, Kramer RA. Environmental Management for Malaria Control: Knowledge and Practices in Mvomero, Tanzania, EcoHealth. 2010;7(4): 507-516

[28] Ukubuiwe AC, Ojianwuna CC, Olayemi IK, Arimoro FO, Ukubuiwe CC. Quantifying the roles of Water pH and Hardness Levels in Development and Biological Fitness in *Culex quinquefasciatus* (Diptera: Culicidae). The Journal of Basic and Applied Zoology. 2020; 81: 5. Doi.org/10.1186/s41936-020-0139-6.

[29] Briegel H. Physiological bases of mosquito ecology. Journal of Vector Ecology. 2003;28:1-11.

[30] Olayemi IK, Ande AT. Life table analysis of *Anopheles gambiae* (Diptera: Culicidae) in relation to malaria transmission. Journal of Vector-borne Diseases. 2009;46: 295-298

[31] Timmermann SE, Briegel H. Larval growth and biosynthesis of reserves in mosquitoes. Journal of Insect Physiology. 1999;45: 461-470.

[32] Ukubuiwe AC, Olayemi IK, Jibrin AI. Genetic Variations in Bionomics of *Culex quinquefasciatus* (Diptera: Culicidae) Mosquito Population in Minna, North-central Nigeria. International Journal of Insect

Science. 2016;8:9-15 (doi:10.4137/ijis.s32516).

[33] Blackmore MS, Lord CC. The relationship between size and fecundity in *Aedes albopictus*. Journal of Vector Ecology. 2000; 25: 212-217.

[34] Armbruster P, Hutchinson RA. Pupal mass and wing length as indicators of fecundity in *Aedes albopictus* and *Aedes geniculatus* (Diptera: Culicidae). Journal of Medical Entomology. 2002;39:699-704.

[35] Ameneshewa B, Service MW. The relationship between female body size and survival rate of the malaria vector *Anopheles arabiensis* in Ethiopia. Medical and Veterinary Entomology. 1996;10: 170-172.

[36] Renshaw M, Service MW, Birley MH. Size variation and reproductive success in the mosquito *Aedes cantans*. Medical and Veterinary Entomology. 1994; 8;179-186.

[37] Lyimo EO, Takken W. Effects of adult body size on fecundity and the pre-gravid rate of *Anopheles gambiae* females in Tanzania. Medicine and Veterinary Entomology. 1993;7: 328-332.

[38] Lyimo EO, Koella JC. Relationship between body size of adult *Anopheles gambiae s.l.* and infection with the malaria parasite *Plasmodium falciparum*. 1991; Swiss Tropical Institute Socinstrasse 57, CH-4002 Basel, Switzerland p1.

[39] Markow TA. Evolutionary ecology and developmental stability. Annual Reviews in Entomology. 1995;7:328- 332.

[40] Mpho M, Callaghan A, Holloway GJ. Effects of temperature and genetic stress on life history and fluctuating wing asymmetry in *Culex pipiens* mosquitoes. European Journal of Entomology. 2002;99:405-412

[41] Kaufmann C, Briegel H. Flight performance of the malaria vectors *Anopheles gambiae* and *Anopheles atroparvus*. Journal of Vector Ecology. 2004;29: 140-153.

[42] Telang A, Wells MA. The effect of larval and adult nutrition on successful autogenous egg production by a mosquito. Journal of Insect Physiology. 2004;50:677-685.

[43] Aparna T, Yiping L, Fernando GN, Mark RB. Effects of larval nutrition on the endocrinology of mosquito egg development. The Journal of Experimental Biology. 2006; 209:645-655.

[44] Briegel H, Hefti M, DiMarco E. Lipid metabolism during sequential gonotrophic cycles in large and small female *Aedes aegypti*. Journal of Insect Physiology. 2002;48: 547-554.

[45] Zhou G, Flowers M, Friedrich K, Horton J, Pennington J, Wells MA. Metabolic fate of [14C]-labelled meal protein amino acids in *Aedes aegypti* mosquitoes. Journal of Insect Physiology. 2004;50:337-349.

[46] Craig MH, Snow RW, Le Suer D. A climate-based distribution model of malaria transmission in sub-Saharan Africa. Parasitology Today. 1999;15:105-111.

[47] Healy KB, Dugas E, Fonseca DM. Development of a degree-day model to predict egg hatch of *Aedes albopictus*. Journal of the American Mosquito Control Association. 2019; 35(4):249-257.

[48] Olayemi IK, Victoria O, Ukubuiwe AC, Jibrin AI. Effects of Temperature Stress on Pre-Imaginal Development and Adult Ptero-fitness of the Vector Mosquito, *Culex quinquefasciatus* (Diptera: Culicidae). Journal of Mosquito Research.

2016;6(14): 1-7 (doi: 10.5376/jmr.2016.06.0014).

[49] Ukubuiwe AC, Olayemi IK, Omalu ICJ, Arimoro FO, Baba BM, Ukubuiwe CC, Odeyemi MO, Adeniyi KA. Influence of Rearing-water Temperature on Life Stages' Vectorial Attributes, Distribution and Utilization of Metabolic Reserves in *Culex quinquefasciatus* (Diptera: Culicidae): Implications for Disease Transmission and Vector Control. The Journal of Basic and Applied Zoology. 2018a;79(32): 1-15. https://doi:10.1186/s41936-018-0045-3

[50] Lanciani CA, Le TM. Effect of temperature on the wing length-body weight relationship in *Anopheles quadrimaculatus*. Journal of American Mosquito Control Association. 1995;11:241-243.

[51] Phasomkusolsil S, Lerdthusnee K, Khuntirat B, Kongtak W, Pantuwatana K, Murphy JR. Effect of temperature on laboratory reared *Anopheles dirus* Peyton and Harrison and *Anopheles sawadwongporni* Rattanarithikul and Green. Southeast Asian Journal of Tropical Medicine and Public Health. 2011; 42(1):63-70.

[52] Ukubuiwe AC, Olayemi IK, Arimoro FO, Omalu ICJ, Odeyemi MO, Ukubuiwe CC, Jibrin AI, Samuel MO. Influence of Fluctuating Temperatures on Morphometry of *Culex quinquefasciatus* (Diptera: Culicidae) Mosquito. Asian Journal of Biological Sciences. 2019;12(3):533-542. Doi: 10.3923/ajbs.2019.533.542

[53] Bowler K. Cellular heat injury. Are membranes involved? In Bowler K, Fuller BJ, (Eds.), *Temperature and Animal Cells.* Society of Experimental Biology Symposium, Cambridge. 1987; 157-185.

[54] Martens P, Kovats RS, Nijhof S, de Vries P, Livermore MTJ, McMichael AJ. Climate change and future populations at risk of malaria. Global Environmental Change. 1999 ; 9:89-107

[55] Clements AN. *The Biology of Mosquitoes: Development, Nutrition and Reproduction.* Chapman & Hall, London, 2000; 675.

[56] De Carvalho SC, Martins-Junior AJ, Lima JB, Valle D. Temperature influence on embryonic development of *Anopheles albitarsis* and *Anopheles aquasalis*. MemÛrias do Instituto Oswaldo Cruz. 2002; 97:1117-1120.

[57] Bayoh MN, Lindsay SW. Effect of temperature on the development of the aquatic stages of *Anopheles gambiae* sensu stricto (Diptera: Culicidae). Bulletin of Entomology Research. 2003;93:375-381.

[58] Rueda LM, Patel KJ, Axtell RC, Stinner RE. Temperature-dependent development and survival rates of *Culex quinquefasciatus* and *Aedes aegypti* (Diptera: Culicidae). Journal of Medical Entomology. 1990;27:892-898.

[59] Ribeiro PB, Costa PRP, Loeck AE, Vianna EES, Silveira Jr P. Exigências térmicas de *Culex quinquefasciatus* (Diptera, Culicidae) em Pelotas, Rio Grande do Sul, Brasil. Iheringia, Série Zoologica. 2004;94 :177-180.

[60] Loetti MV, Nora EB, Paula P, Schweigmann N. Effect of temperature on the development time and survival of preimaginal *Culex hepperi* (Diptera: Culicidae). Revolutionary Society of Entomology. 2008; 67 (3-4):79-85.

[61] Alto BW, Juliano SA. Temperature effects on the dynamics of *Aedes albopictus* (Diptera: Culicidae) populations in the laboratory. Journal of Medical Entomology. 2001;38:548-556.

[62] Afrane YA, Zhou G, Lawson BW, Githeko AK, Yan G. Effects of microclimatic changes caused by

deforestation on the survivorship and reproductive fitness of *Anopheles gambiae* in western Kenya highlands. American Journal of Tropical Medicine and Hygiene. 2006;74:772-778.

[63] Joshi D. Effect of fluctuating and constant temperatures on development, adult longevity and fecundity in the mosquito *Aedes krombeini*. Journal of Thermal Biology. 1996;21:151-154.

[64] Dodson BL, Kramer LD, Rasgon JL. Effects of larval rearing temperature on immature development and West Nile Virus vector competence of *Culex tarsalis*. Parasites and Vectors. 2012; 5:199

[65] Reisen WK. Effect of temperature on *Culex tarsalis* (Diptera: Culicidae) from the Coachella and San Joaquin Valleys of California. Journal of Medical Entomology. 1995; 32:636-645.

[66] Beier MS, Beier JC, Merdan AA, el Sawaf BM, Kadder MA. Laboratory rearing techniques and adult life table parameters for *Anopheles sergentii* from Egypt. Journal of American Mosquito Control Association. 1987;3:266-270.

[67] Loetti MV, Burroni NE, Schweigmann N, de Garin A. Effect of different thermal conditions on the pre-imaginal biology of *Culex apicinus* (Philippi, 1865) (Diptera: Culicidae). Journal of Vector Ecology. 2007;32: 106.

[68] Le Sueur D, Sharp BL. Temperature-dependent variation in *Anopheles merus* larval head capsule width and adult wing length: implications for anopheline taxonomy. Medical and Veterinary Entomology. 1991; 5:55-62.

[69] Patrick ML, Ferreira RL, Gonzalez RJ, Wood CM, Wilson RW, Bradley TJ, Val AL. Ion regulatory patterns of mosquito larvae collected from breeding sites in the Amazon rain forest. Physiology, Biochemistry and Zoology. 2002; 75: 215-222.

[70] Gillott C. *Entomology*. 3rd Ed. Springer publishing. 2005; 500-511.

[71] Patrick ML, Gonzalez RJ, Wood CM, Wilson RW, Bradley TJ, Val AL. The characterization of ion regulation in Amazonian mosquito larvae: Evidence of phenotypic plasticity, population based disparity, and novel mechanisms of ion uptake. Physiology, Biochemistry and Zoology. 2002; 75: 223-236.

[72] Minakawa N, Sonye G, Yan G. Relationships between occurrence of *Anopheles gambiae* s. l. (Diptera: Culicidae) and size and stability of larval habitats. Journal of Medical Entomology. 2005;42 (3): 295-300.

[73] Yee DA, Kneitel JM, Juliano SA. Environmental Correlates of Abundances of Mosquito Species and Stages in Discarded Vehicle Tires. Journal of Medical Entomology. 2010;47(1): 53-62.

[74] Clark TM, Flis BJ, Remold SK. pH tolerances and regulatory abilities of freshwater and euryhaline Aedine mosquito larvae. The Journal of Experimental Biology. 2004;207: 2297-2304.

[75] Locke M. The Wigglesworth lecture: Insects for studying fundamental problems in biology. Journal of Insect Physiology. 2001;47:495-507.

[76] Nadia K, Thamer M, Saad SA. The Effect of Different NaCl, pH level of Survival of *Culex* species (Diptera: Culicidae) Larvae in Basrah. Journal of Basrah Research (Science). 2005;3(2): 31-36.

[77] Singare PU, Lokhande RS, Pathak PP. Soil pollution along Kalwa Bridge at Thane Creek of Maharashtra, India. Journal of Environmental Protection. 2010;1:121-128.

[78] Woodhill AR. A comparison of factors affecting the development of

three species of mosquitoes, etc. Proceedings of Limnological Society. N. S. W. 1942; 67:95-97.

[79] Buchman NW. Untersuchungenti ber die Bedeutungd er Wasserst offionenkonzentration fur die Entwicklung der Mtickenlarven. 2. Angew. Entomology. 1931;18:404-416.

[80] Robert V, Awono-Ambene HP, Thiolouse J. Ecology of larval Mosquitoes (Diptera: Culicidae) in market-garden wells in urban Dakar, Senegal. Journal of Medical Entomology. 1998; 35(6):948-955.

[81] World Health Organisation (1997). Guidelines for drinking water quality (2nd Edition). Volume 3. Surveillance and Control of community supplies. Geneva, Switzerland.

[82] Molokwu CN, Okpokwasili GC. The Effect of Water hardness on Egg Hatchability and Larval Viability of *Clarias gariepinus*. Aquaculture International. 2004;10:57-64.

[83] Blanksma CB, Eguia KL, Lazorchak JM, Smith ME, Wratschko M, Schoenfuss HL. Effects of water Hardness on Skeletal Development and Growth in Juvenile fathead minnows. Aquaculture. 2009; 286;226-232.

[84] Water Quality Association, WQA. Scale deposits are a typical indicator of hard water. www.wqa.org, 2018 (Accessed 14/01/2018).

[85] Poteat MD, Buchwalter DB. Calcium uptake in aquatic insects: influences of phylogeny and metals (Cd and Zn). The Journal of Experimental Biology, 2014;217:1180-1186 doi:10.1242/jeb.097261

[86] Day JF, Van-Handel E. Differences between the nutritional reserves of laboratory-maintained and field-collected adult mosquitoes. Journal of American Mosquito Control Association. 1986; 2(2):154-157.

[87] Briegel H. Fecundity, metabolism, and body size in *Anopheles* (Diptera: Culicidae), vectors of malaria. Journal of Medical Entomology. 1990;27:839-850.

[88] Kant R, Pandey SD, Sharma SK. Mosquito breeding in relation to aquatic vegetation and some physico-chemical parameters in rice fields of central Gujarat. Indian Journal of Malariology. 1996;33:30-40.

[89] Piyaratne MK, Amerasinghe FP, Amerasinghe PH, Konradsen F. Physico – chemical characteristics of *Anopheles culicifacies* and *Anopheles varuna* breeding water in a dry zone stream in Sri Lanka. Journal of Vector Borne Diseases. 2005;42;61 – 67.

[90] Mwangangi JM, Mbogo CM, Muturi EJ, Nzovu JG, Kabiru EW, Githure JI, Novak RJ, Beier JC. Influence of biological and physicochemical characteristics of larval habitats on the body size of *Anopheles gambiae* mosquitoes (Diptera: Culicidae) along the Kenyan coast. Journal of Vector Borne Diseases. 2007;44:122-127.

[91] Oyewole IO, Momoh OO, Anyasor GN, Ogunnowo AA, Ibidapo CA, Oduola OA, Obansa JB, Awolola TS. Physico-chemical characteristics of *Anopheles* breeding sites: Impact on fecundity and progeny development. African Journal of Environmental Science and Technology. 2009;3(12):447-452.

[92] Olayemi IK, Omalu ICJ, Famotele OI, Shegna SP, Idris B. Distribution of mosquito larvae in relation to physico-chemical characteristics of breeding habitats in Minna, North Central Nigeria. Reviews in Infection. 2010;1(1): 49-53.

[93] Mgbemena IC, Ebe T. Distribution and Occurrence of Mosquito Species in

the Municipal Areas of Imo State, Nigeria. Analele UniversităŃii din Oradea - Fascicula Biologie. 2012;19(2):93-100.

[94] Ukubuiwe AC, Olayemi IK, Omalu ICJ, Arimoro FO, Ukubuiwe CC. Morphometric Diagnosis of the Effects of Water Hardness on Development of Immature Life Stages and Adult Vectorial Fitness of *Culex quinquefasciatus* (Diptera: Culicidae) Mosquito. Zoomorphology. 2018;137(4): 511-518. https://doi.org/10.1007/s00435-018-0415-x

[95] Aminuwa H, Olayemi IK, Ukubuiwe AC, Adeniyi KA, Odeyemi MO. Evaluation of Critical Larval Habitat Physico-chemical Factors on Embryonic Development and Adult Fitness of *Culex quinquefasciatus* mosquitoes (Diptera: Culicidae). Malaya Journal of Bioscience. 2018;5(2): 48-56.

[96] Chocorosqui VR, Panizzi AR. Photoperiod influence on the Biology and phenological characteristics of *Dichelops melacanthus* (Dallas, 1851) (Heteroptera: Pentatomidae), Brazilian Journal of Biology. 2003;63(4):655-664.

[97] MacRae TH. Diapause: diverse states of developmental and metabolic arrest. Journal of Biological Research. 2005;3:3-14.

[98] Cloutier EJ, Beck SD, McLeod DGR, Silhacek DL. Neural transplants and insect diapause. Nature. 1962;135:1222-1224.

[99] Mathias D, Laura KR, William EB, Holzapfel CM. Evolutionary Divergence of Circadian and Photoperiodic Phenotypes in the Pitcher-Plant Mosquito, *Wyeomyia smithii*. Journal of Biological Rhythms. 2006;21(2):132-139.

[100] Reznik SY, Vaghina NP. Photoperiodic control of development and reproduction in *Harmonia axyridis* (Coleoptera: Coccinellidae). European Journal of Entomology. 2011;108:385-390.

[101] Śniegula S, Nilsson-Örtman V, Johansson F. Growth Pattern Responses to Photoperiod across Latitudes in a Northern Damselfly. PLoS ONE. 2012; 7(9):e46024. doi: 10.1371/journal.pone.0046024

[102] Adkisson PL. Action of the Photoperiod in Controlling Insect Diapause. The American Naturalist. 1964;98(902):357-374.

[103] MacRae TH. Gene expression, metabolic regulation and stress tolerance during diapause. Cellular and Molecular Life Sciences. 2010;67(14):2405-2424.

[104] Urbaneja A, Llacer E, Garrido A, Jacas J. Effect of variable photoperiod on development and Survival of *Cirrospilus* sp. Nr. *Lyncus* (Hymenoptera: Eulophidae), An Ectoparasitoid of *Phyllocnistis citrella* (Lepidoptera: Gracillariidae). Florida Entomologist. 2001;84(2);305-307.

[105] Oda T, Nuorteva, P. Autumnal photoperiod and the development of follicles in *Culex pipiens pipiens* L. (Diptera, Culicidae) in Finland. Annals of Entomology Fennici. 1987; 53:33-35.

[106] Lanciani CA, Anderson JF. Effect of photoperiod on longevity and metabolic rate in *Anopheles quadrimaculatus*. Journal of America Mosquito Control Association. 1993;9:158-163.

[107] Vinogradova EB, Karpova SG. Effect of photoperiod and temperature on the autogeny rate, fecundity and wing length in the urban mosquito, *Culex pipiens pipiens f. molestus* (Diptera, Culicidae). International Journal of Dipteran Research. 2006;17 (1):3-12.

[108] Bradshaw WE, Holzapfel CM. Biology of tree-hole mosquitoes:

photoperiodic control of development in northern *Toxorhynchites rutilus* (Coq.). Canadian Journal of Zoology. 1975;53:889-893.

[109] Carmine LA, Ronald E. Effect of Photoperiods on *Anopheles quadrimaculatus*. Florida Entomologist. 1993;76(4):622.

[110] Lanciani CA. Photoperiod and longevity in *Anopheles crucians*. Journal of America Mosquito Control Association, 1993;9:308-312.

[111] Ukubuiwe AC, Olayemi IK, Omalu ICJ, Arimoro FO, Baba BM, Ukubuiwe CC. Effects of Varying Photoperiodic Regimens on Critical Biological Fitness traits of *Culex quinquefasciatus* (Diptera: Culicidae) Mosquito Vector. International Journal of Insect Science. 2018;10: 1-10. 10:1179543318767915 (doi: 10.1177/1179543318767915)

[112] Ukubuiwe AC, Olayemi IK, Omalu ICJ, Arimoro FO, Baba BM, Ukubuiwe CC. Influence of Variable Photoperiod on Life-stages Mobilization of Teneral Reserves in *Culex quinquefasciatus* (Diptera: Culicidae): Implication for Environmental Manipulation for Vector Control. Molecular Entomology. 2018;9(1): 1-10 (doi: 10.5376/me.2018.09.0001)

[113] Jesha MM, Sebastian NM, Sheela PH, Mohamed IS, Manu AY. Mosquito Density in Urban Kerala: A Study to Calculate Larval Indices in Municipal Area of Perinthalmanna. Indian Journal of Forensic and Community Medicine. 2015;2(1):7-12.

[114] Tsurim I, Silberbush A, Ovadia O, Blaustein L, Margalith Y. Inter- and Intra-Specific Density-Dependent Effects on Life History and Development Strategies of Larval Mosquitoes. PlosOne. 2013;8:e57875.

[115] Silberbush A, Tsurim I, Rosen R, Margalith Y, Ovadia O. Species-Specific Non-Physical Interference Competition among Mosquito Larvae. PLoSONE. 2014;9(2):e88650. doi: 10.1371/journal.pone.008865.

[116] Schneider P, Takken W, McCall PJ. Interspecific competition between sibling species larvae of *Anopheles arabiensis* and *An. gambiae*. Medical and Veterinary Entomology. 2000; 14:165-170.

[117] Suleman M. The Effects of Intraspecific Competition for Food and Space on the Larval Development of *Culex quinquefasciatus*. Mosquito News. 1982;42:347-355.

[118] Timmermann SE, Briegel H. Water depth and larval density affect development and accumulation of reserves in laboratory populations of mosquitoes. Bulletin of the Society of Vector Ecology. 1993;18:174-187.

[119] Reisen WK, Emory RW. Intraspecific competition in *Anopheles stephensi* (Diptera Culicidae) II. The effects of more crowded densities and the addition of antibiotics. Canadian Entomology. 1977; 109:1475-1480.

[120] Okwa OO, Rasheed A, Adeyemi A, Omoyeni M, Oni L, Fayemi A, Ogunwomoju A. *Anopheles* species abundances, composition and vectorial competence in six areas of Lagos: Nigeria. Journal of Cell and Animal Biology. 2007;1(2):19-23

[121] Gimnig JE, Ombok M, Otieno S, Kaufman MG, Vulule JM, Walker ED. Density-dependent development of *Anopheles gambiae* (Diptera: Culicidae) larvae in artificial habitats. Journal of Medical Entomology. 2002;39:162-172.

[122] Ye-Ebiyo Y, Pollack RJ, Kiszewski A, Spielman A. Enhancement of development of larval *Anopheles arabiensis* by proximity to flowering maize (*Zea mays*) in turbid water and when crowded. American Journal of

Tropical Medicine and Hygiene. 2003;68:748-752.

[123] Tseng M. Sex-specific response of a mosquito to parasites and crowding. Proceedings of the Royal Society B-Biological Sciences. 2004;271:186-188.

[124] Reisen WK. Intraspecific competition in *Anopheles stephensi* Liston. Mosquito News. 1975;35, 473-482

[125] Bedhomme S, Agnew P, Sidobre C, Michalakis Y. Sex specific reaction norms to intraspecific larval competition in the mosquito *Aedes aegypti*. Journal of Evolutionary Biology. 2003;16:721-730.

[126] Ukubuiwe AC, Ojianwuna CC, Olayemi IK, Arimoro FO, Omalu ICJ, Ukubuiwe CC, Baba BM. Quantifying the Influence of Larval density on Disease Transmission Indices in *Culex quinquefasciatus,* the major African vector of filariasis. International Journal of Insect Science. 2019;11:1-11. Doi: 10.1177/1179543319856022

[127] Briegel H. Metabolic relationship between female body size, reserves, and fecundity of *Aedes aegypti*. Journal of Insect Physiology. 1990;36:165-172.

[128] Gleiser RM, Urrutia J, Gorla DE. Effects of crowding on populations of *Aedes albifasciatus* larvae under laboratory conditions. Entomologia Experimentalis et Applicata. 2000;95:135-140.

[129] Legros M, Lloyd AL, Huang YX, Gould F. Density-dependent intraspecific competition in the larval stage of *Aedes aegypti* (Diptera: Culicidae): revisiting the current paradigm. Journal of Medical Entomology. 2009;46: 409-419.

[130] Hawley WA. The effect of larval density on adult longevity of a mosquito, *Aedes sierrensis*: epidemiological consequences. Journal of Animal Ecology. 1985;54:955-964.

[131] Takken W, Klowden MJ, Chambers GM. Effect of body size on host seeking and blood meal utilization in *Anopheles gambiae* sensu stricto (Diptera: Culicidae): the disadvantage of being small. Journal of Medical Entomology. 1998;35:639-645.

[132] Macia´ A. Effects of larval crowding on development time, survival and weight at metamorphosis in *Aedes aegypti* (Diptera: Culicidae). Revista de la Sociedad Entomolo´gica Argentina. 2009;68:107-114.

[133] Roberts D, Kokkinn M. Larval crowding effects on the mosquito *Culex quinquefasciatus*: physical or chemical? Entomologia Experimentalis et Applicata. 2010; 135:271-275.

[134] Rajagopalan PK, Yasuno M, Menon PK. Density effect on survival of immature stages of *Culex pipiens fatigans* in breeding sites in Delhi villages. The Indian Journal of Medical Research. 1976 ;64:688-708.

[135] Mori A. Effects of larval density and nutrition on some attribute of immature and adult *Ae. albopictus*. Tropical Medicine. 1979;21:85-103.

[136] Reisen WK, Milby MM, Bock ME. The effects of immature stress on selected events in the life history of *Culex tarsalis*. Mosquito News. 1984;44:385-395.

[137] Bradshaw WE, Holzapfel CM. Reproductive consequences of density-dependent size variation in the pitcher plant mosquito, *Wyeomyia smithii* (Diptera: Culicidae). Annals of Entomological Society of America. 1992;85: 274-281

[138] Barbosa P, Peters TM. Some effects of overcrowding on the respiration of larval *Aedes aegypti*. Entomologia

Experimentalis et Applicata. 1973;16: 146-156.

[139] Muriu SM, Coulson T, Mbogo CM, Godfray HCJ. Larval density dependence in *Anopheles gambiae* s.s., the major African vector of malaria. Journal of Animal Ecology. 2013;82:166-174.

[140] Roberts D. Overcrowding of *Culex sitiens* (Diptera: Culicidae) larvae: population regulation by chemical factors or mechanical interference. Journal of Medical Entomology. 1998;35:665-669.